T0280433

Data-Centric Systems and Applications

Intelligent data management is the backbone of all information processing and has hence been one of the core topics in computer science from its very start. This series is intended to offer an international platform for the timely publication of all topics relevant to the development of data-centric systems and applications. All books show a strong practical or application relevance as well as a thorough scientific basis. They are therefore of particular interest to both researchers and professionals wishing to acquire detailed knowledge about concepts of which they need to make intelligent use when designing advanced solutions for their own problems.

Special emphasis is laid upon:

- Scientifically solid and detailed explanations of practically relevant concepts and techniques
  (what does it do)
- Detailed explanations of the practical relevance and importance of concepts and techniques
  (why do we need it)
- Detailed explanation of gaps between theory and practice
  (why it does not work)

According to this focus of the series, submissions of advanced textbooks or books for advanced professional use are encouraged; these should preferably be authored books or monographs, but coherently edited, multi-author books are also envisaged (e.g. for emerging topics). On the other hand, overly technical topics (like physical data access, data compression etc.), latest research results that still need validation through the research community, or mostly product-related information for practitioners ("how to use Oracle 9i efficiently") are not encouraged.

More information about this series at http://www.springer.com/series/5258

Antonio Badia

SQL for Data Science

Data Cleaning, Wrangling and Analytics with Relational Databases

Antonio Badia
Computer Engineering & Computer Science
University of Louisville
Louisville, KY, USA

ISSN 2197-9723 ISSN 2197-974X (electronic)
Data-Centric Systems and Applications
ISBN 978-3-030-57591-5 ISBN 978-3-030-57592-2 (eBook)
https://doi.org/10.1007/978-3-030-57592-2

This Springer imprint is published by the registered company Springer Nature Switzerland AG.
The registered company address is: Gewerbestrasse 11, 6330 Cham, Switzerland

Preface

Data Science (or Data Analytics, or whatever one prefers to call it) is a 'hot' topic right now. There is an explosion of courses on the subject, especially online: many universities and several for-profit and non-profit organizations (Coursera, edX, Udacity, Udemy, DataCamp, and many others) offer on-campus and online courses, certification, and degrees. The coverage of these offerings is quite diverse, reflecting the fact that Data Science is still a young and evolving field. However, many courses seem to coalesce around a few topics (Machine Learning, mostly) and tools (R, Python, and SQL, mostly). What few of these courses offer is a textbook.

There are already many books on databases and SQL, but almost all of them focus on the traditional curriculum for Computer Science majors or Information Systems majors (there are a few exceptions, like [11] and [17]). In contrast, the present book explains SQL *within the context* of Data Science and is more in line with what is being taught in these new courses. This book introduces the different parts of SQL as they are needed for the tasks usually carried out during data analysis. Using the framework of the *data life cycle*, it focuses on the steps that are given the short shift in traditional textbooks, like data loading, cleaning, and pre-processing.

This book is for anyone interested in Data Science and/or databases. It should prove useful to anyone taking any of the abovementioned courses, online or on-campus, as well as to students working on their own. It assumes very little from the reader; it just demands a bit of 'computer fluency,' but no background on databases or data analysis. In general, all concepts are introduced intuitively and with a minimum of specialized jargon. It contains an appendix (Appendix A) meant to help students without prior experience with databases, with instructions on how to download and install the two open-source database systems (MySQL and Postgres) that we use for examples throughout the book. All readers of the book are encouraged to install both systems and follow the book along with a computer in order to practice, do the exercises, and play around—simply reading the book alone is going to be much less useful than *using* it.

The book is organized as follows: Chapter 1 describes the *Data Life Cycle*, the sequence of stages, from data acquisition and ingestion until archiving, that data goes through as it is prepped for analysis and then actually analyzed, together with the different activities that take place at each stage. It also explains the different ways that datasets can be organized, and the different types of data one may have to deal with. Many students have an intuitive understanding of the concepts in this chapter, but it is useful to have it all together in one place and to give a name to each concept for later reference. Chapter 2 gets into databases proper, explaining how relational databases organize data. The chapter also explains how data in tables *should* look like (what Hadley Wickham has called *tidy* data [19]), a point which is not traditionally emphasized and can lead to severe problems down the road. Non-traditional data, like XML and text, are also covered. Chapter 3 introduces SQL *queries*, the SQL commands that allow us to ask questions about the data. Unlike traditional textbooks, queries and their parts are described around typical data analysis tasks (data exploration, cleaning, and transformation). These tasks are vital for a proper examination of the data but are frequently overlooked in Data Mining and Machine Learning textbooks. Chapter 4 introduces some basic techniques for Data Analysis. Even though this is not the focus of the book, the chapter shows that SQL can be used for some simple analyses without too much complication.

After this part, which constitutes the core of the book, Chap. 5 introduces additional SQL constructs that come in handy in a variety of situations. This chapter completes the coverage of SQL queries so that readers get an overview of all the main aspects of this important topic. Chapter 6 briefly explains how to use SQL from within R and from within Python programs. This chapter is not an introduction to R (or to Python) and, unlike other chapters in the book, does assume that the reader is already familiar with at least the basics of R and Python. It focuses on how these languages can interact with a database, and how what has been learned about SQL can be leveraged to make life easier when using R or Python.

The book also contains another appendix (besides the one already mentioned), which introduces some basic approaches for handling very large datasets. The purpose of this appendix is to demystify the ideas behind the vague label *Big Data* and give the readers basic guidance on how to use their newly acquired skills in this world.

As in many textbooks, none of what this one contains is new. This book covers the same (or very similar) content to what can be found in many sources, especially online. What this book does is to put it all together under one roof and to give it some order and structure. In many blogs and sites, the material is presented as an answer to a particular question (how do you...?), which may be useful to someone with a specific need but gives the impression that learning SQL is about a bag of tricks. Here, the material is logically organized using the idea of the data life cycle so that all the concepts introduced can be understood as parts of a coherent whole.

Data Science itself is a relatively new and still changing field, but it has deep roots, as it uses approaches and techniques from well-established fields, mostly math (statistics, linear algebra, and others) and computer science (databases, machine learning, and others). As a result, the same concept is sometimes given different names by different authors in different textbooks. Whenever I am aware of this, I have given a list of known names so that readers with different backgrounds can relate what is in here with what they already know.

The goal of the book is to introduce some basic concepts to a wide variety of readers and provide them a good foundation on which they can build. After going through this book, readers should be able to profitably learn more about Data Mining, Machine Learning, and database management from more advanced textbooks and courses. It is my hope that most of them feel that they have been given a springboard from which they are in a good position to dive deeper into the fascinating world of data analysis.

Louisville, KY, USA
July 2020

Antonio Badia

Contents

Chapter 1
The Data Life Cycle

It is sometimes said that "data is the new oil." This is true in several ways: in particular, data, like oil, needs to be processed before it is useful. Crude oil undergoes a complex refining procedure as the substance that comes out of wells is transformed into several products, mostly fuels (but also many other useful by-products, from asphalt to wax). A complex infrastructure, from pipelines to refineries, supports this process. In a similar way, raw data must be thoroughly treated before it can be used for anything. Unfortunately, there is not a big and sophisticated infrastructure to support data processing. There are many tools that support some of the steps in the process, but it is still up to every practitioner to learn them and combine them appropriately.

In this chapter, we introduce the stages through which data passes as it is refined, analyzed, and finally disposed of. The collection of stages is usually called the *data life cycle*, inspired by the idea that data is 'born' when it is captured or generated and goes through several stages until it reaches 'maturity' (is ready for analysis) and finally an end-of-life, at which point it is deleted or archived. Data analysis, which is the focus of Data Mining and Machine Learning books and courses, is but one step in this process. The other steps are equally important and often neglected.

The main purpose of this chapter is to introduce a framework that will help organize the contents of the rest of the book. As part of this, it introduces some basic concepts and terms that are used in the following chapters. In particular, it provides a classification of the most common types of *datasets* and *data domains* that will be useful for later work. We will come back to these topics throughout the book, so the reader is well served to start here, even though SQL itself does not appear until the next chapter. Also, for readers who are new to data analysis, this chapter provides a basic outline of the field.

A. Badia, *SQL for Data Science*, Data-Centric Systems and Applications,
https://doi.org/10.1007/978-3-030-57592-2_1

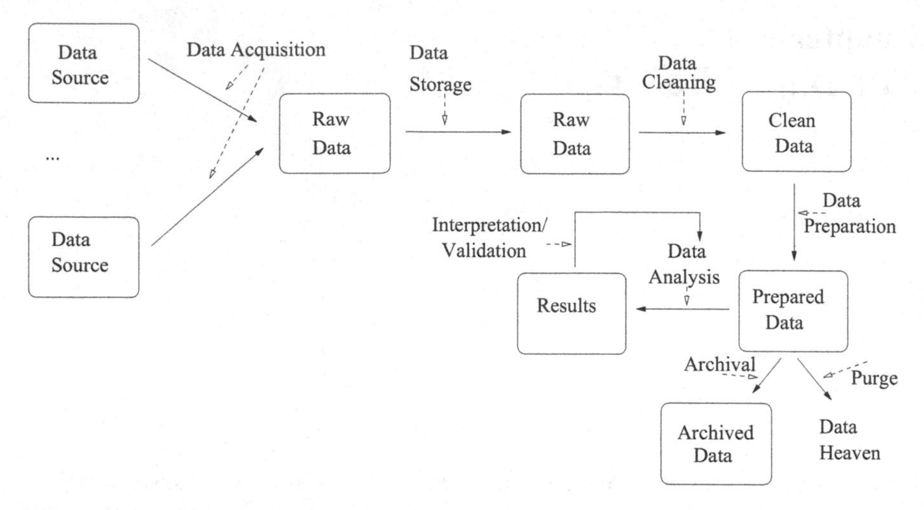

Fig. 1.1 The data life cycle

1.1 Stages and Operations in the Data Life Cycle

The term *data life cycle* refers both to the transformations applied to data and to the states that data goes through as a result of these transformations. While there is not, unfortunately, general agreement on the exact details of what is involved at each transformation and state, or how to refer to them, there is a wide consensus on the basic outlines. The states of the cycle can be summarized as follows:

Raw data → cleaned data → prepared data → data + results → archived data

The arrows here indicate precedence; that is, raw data comes first, and cleaned data is extracted from it, and so on. The activities are usually described as follows:[1]

Data Acquisition/capture → data storage → data cleaning/wrangling/enrichment → data analysis → data archival/preservation

Again, the arrows indicate precedence; data acquisition/capture happens first, followed by data storage, and so on.

The diagram in Fig. 1.1 shows the activities and stages together. We now describe each part in more detail.

The first activity in data analytics is to acquire, collect, or gather data. This happens in different ways. Sometimes existing sources of data are known and

[1] Some activities are given different names in different contexts.

accessible, sometimes a prior step that uncovers sources of relevant data[2] must be carried out. What we obtain as the result of this step is called *raw data*.

It is very important to understand that "raw" refers to the fact that this is the data before any processing has been applied to it, but does not indicate that this data is "neutral" or "unfiltered." In statistics, the domain of study is called the *population*, and the data collected about the domain is called the *sample*. It is understood that the sample is always a subset of the whole population and may vary in size from a very small part to a substantial one. However, the sample is never the population, and the fact that sometimes we have a large amount of data should not fool us into believing otherwise. For analysis of the sample to provide information about the population, the sample must be *representative* of the population. For this to happen, the sample must be chosen at random from elements of the population which are equally likely to be selected. It is very typical in data science that the data is collected in an *opportunistic* manner, i.e. data is collected because it is (easily) available. Furthermore, in science data usually comes from experiments, i.e. a setting where certain features are controlled, while a lot of data currently collected is *observational*, i.e. derived from uncontrolled settings. There are always some decisions as to what/when/how to collect data. Thus, raw data should not be considered as an absolute source of truth, but carefully analyzed.

When data comes in, we can have two different situations. Sometimes datasets come with a description of the data they contain; this description is called *metadata* (metadata is described in some detail in Sect. 1.4). Sometimes the dataset comes without any indication of what the data is about, or a very poor one. In either situation, the first step to take is *Exploratory Data Analysis (EDA)* (also called *data profiling*). In this step, we try to learn the basic characteristics of the data and whatever objects or events or observations it describes. If there is metadata, we check the dataset against it, trying to validate what we have been told—and augment it, if possible. If there is not metadata, this is the moment to start gathering it. This is a crucial step, as it will help us build our understanding of the data and guide further work. This step involves activities like classifying the dataset, getting an idea of the attributes involved, and for each attribute, getting an idea of data distribution through *visualization* techniques, or *descriptive statistics* tools, like histograms and measures of centrality or dispersion.[3]

We use the knowledge gained in EDA to determine whether data is correct and complete, at least for current purposes. Most of the time, it will not be, so once we have determined what problems the data has, we try to fix them. There are often issues that need to be dealt with: the data may contain errors or omissions, or it may not be in the right format for analysis. There are many *sources* of errors: manual (unreliable) data entry; changes in layout (for records); variations in measurement, scale, or format (for values); changes in how default or missing values are marked; or outdated values (called "gaps" in time series). Many of these issues can only be

[2] This step is referred to as *source discovery*.

[3] Readers not familiar with these notions will be introduced to the basics in Chap. 3.

addressed by changing the data gathering or acquisition phase, while others have to be fixed once data is acquired.[4] The tools and techniques used to fix these problems are usually called *data cleaning* (or *data cleansing*, *data wrangling*, *data munging*, among others). The issues faced, and the typical operations used, include

- Finding and handling *missing values*. Such values may be explicitly or implicitly denoted. Explicitly denoted missing values are usually identified with a marker like 'NULL,' 'NA' (or "N.A.," for "Not Available") or similar; but different datasets may use different conventions. Implicit missing values are denoted by the absence of a value instead of by a marker. Because of this variety, finding missing values is not always easy. Handling the absence of values can be accomplished simply by deleting incomplete data, but there are also several techniques to *impute* a missing value, using other related values in the dataset. For example, assume that we have a dataset describing people, including their weight in pounds. We realize that sometimes the weight is missing. We could look for the weight of people with similar age, height, etc. in the dataset and use such values to fill in for the missing ones.
- Finding and handling *outliers*. Outliers are data values that have characteristics that are very different from the characteristics of most other data values in a set. For example, assume that in the people dataset we also have their height in feet. This is a value that usually lies in the 4.5–6.5 range; anyone below or above is considered very short or very tall. A value of 7.5 is possible, but suspicious; it could be the result of an error in measurement or data entry. As this example shows, finding outliers (and determining when an outlier is a legitimate value or an error) may be context-dependent and extremely hard.
- Finding and handling *duplicate data*. When two pieces of the dataset refer to the same real-world item (entity, fact, event, or observation), we say the data *contains duplicates*. We usually want to get rid of duplicate data, since it could bias (or otherwise negatively influence) the analysis. Just like dealing with outliers, this is also a complex task, since it is usually very hard to come up with ways to determine when duplicate data exists. Using again the example of the people dataset, it is probably not smart to assume that two records with the same name refer to the same person; some names are very common and we could have two people that happen to share the same name. Perhaps if two records have the same name and address, that would do—although we can imagine cases where this rule does not work, like a mother and a daughter with the same name living together. Maybe name, address, and age will work? Many times, the possibility of duplication depends on the context; for instance, if our dataset comes from children in a certain school, first and last name and age will usually do to determine duplication; but if the dataset comes from a whole city, this may not be enough.

[4]The overall management of issues in data is sometimes called *Data Quality*; see Sect. 1.4.

The result of these activities is usually referred to as *clean data*, as in 'data that has been cleaned and fixed.' While cleaning the data is a necessary pre-requisite for any type of analysis, at this point the data is still not ready to be analyzed. This is because different types of analysis may require different additional treatment. Therefore, another step, usually called *data pre-processing* or *data preparation* is carried out in order to prepare the data for analysis. Typical tasks of this step include:

- Transformations to put data values in a certain format or within a certain frame of reference. This involves operations like *normalization, scaling*, or *standardization*.[5]
- Transformations that change the data value from one type to another, like *discretization* or *binarization*.
- Transformations that change the structure of the dataset, like *pivoting* or *(de)normalization*. Most data analysis tools assume that datasets are organized in a certain format, called *tabular data*; datasets not in this format need to be restructured. We describe tabular data in the next section and discuss how to restructure datasets in Sect. 3.4.

Data is now finally ready for analysis. Many techniques have been developed for this step, mostly under the rubric of Statistics, Data Mining, and Machine Learning. These techniques are explained in detail in many other books and courses; in this book we explain a selected few in detail (including an implementation in SQL) in Chap. 4.

Once data has been analyzed, the results of the analysis are usually examined to see if they confirm or disprove any hypothesis that the researcher/investigator may have in mind. The results sometimes generate further questions and produce a cycle of further (or alternative) data analysis. They can also force a rethinking of assumptions and may lead to alternative ways of pre-processing the data. This is why there is a *loop* in Fig. 1.1, indicating that this may become an *iterative* process.

Finally, once the cycle of analysis is considered complete, the results themselves are stored, and a decision must be taken about the data. The data is either *purged* (deleted) or *archived*, that is, stored in some long-term storage system in case it is useful in the future. In many cases, the data is *published* so it can be shared with other researchers. This enables others to reproduce an analysis, to make sure that the results obtained are correct. The publication also allows the data to be reused for different analyses. Whenever data is published, it is very important that it be accompanied by its metadata, so that others can understand the meaning of the dataset (what exactly it is describing) as well as its scope and limitations. If the data was cleaned and pre-processed, those activities should also be part of the metadata. In any case, data (like oil) should be disposed of carefully.

[5] Again, readers not familiar with these should wait until their introduction in the next chapter.

1.2 Types of Datasets

Our first task is to understand the data. Here we describe how datasets are usually classified and described.

There are, roughly speaking, two very different types of data: *alphanumeric* and *multimedia* data. Multimedia refers to data that represents audiovisual (video, images, audio) information. This data is usually encoded using one of the several standards for such media (for instance, JPEG for digital images[6] or MPEG for audio/video[7]). Alphanumeric data refers to collections of characters[8] used to represent alphabetic (names) and numeric individual datum. For instance, '123' represents a number (an integer) in decimal notation; 'blue' represents the name of a color. Such data is used to provide basic values, which are then grouped or organized in several ways (described below). Most methods for data analysis have been developed to deal with alphanumeric data, and that is the only data that we cover in this book. Handling multimedia data requires specialized tools: in order to display the image or play the video or music, a special program (a 'video/audio player') that understands how the encoding works is needed.[9]

An alphanumeric dataset (henceforth, simply a 'dataset') is a collection of *data items*. An item is usually called a *row* or *tuple* in database parlance; a *record*, in general Computer Science parlance; an *observation*, in statistical parlance; or an *entity*, *instance*, or a *(data) point* in other contexts. Each item describes a real-world entity, fact, or event; it consists of a group of related characteristics, each one giving information about some aspect of the entity, fact, or event being described. Such characteristics are called *attributes* in database parlance; *variables*, in Statistics; *features* in Machine Learning; and *properties* or *measurements* in other contexts. For instance, in our previous example of the people dataset, we implicitly assumed that the data was an assemblage of items, each item describing a person, and that the items were composed of attributes describing (among others) the name, address, age, weight, and height of each person. **Important note:** in this book we will use the terms *record* (although we will still use *row* for data in tables) and *attribute* from now on as unifying vocabulary; in formulas, we will use $r, s, r_1, r_2, \ldots$ as variables over records and $A, B, A_1, A_2, \ldots$ as variables over attributes.

The number of records in the dataset is usually termed its *size* (in databases, the *cardinality*). Conversely, the number of attributes present in a record is called the *dimensionality*. In some cases, all records in a dataset are similar and share the same (or almost the same) dimensionality, so that we can speak of the dimensionality of the dataset too.

[6]https://en.wikipedia.org/wiki/Image_file_formats.

[7]https://en.wikipedia.org/wiki/MPEG-1.

[8]In this context, a character is any symbol that can be produced by a key on the computer's keyboard: letters, digits, punctuation marks, and so on.

[9]There are books describing such tools and methods, although they tend to be quite technical.

Depending on the exact nature of the records, datasets can be classified as *structured*, *semistructured*, and *unstructured*. In this book, we will focus mainly on structured (also called 'tabular') data, because this is the kind of data that relational databases handle best, but also because this is the kind of data that is most commonly assumed when talking about data analysis, and the one targeted by most techniques. However, relational databases are also perfectly capable of handling semistructured and unstructured data, and we will also cover analysis of this data later in the book. Therefore, we start with a description of each kind of data.

The attributes that make up a data record or record are called the *schema* of the record. The value of an attribute may be a number or a label, in which case it is called *simple*, or it may have a complex structure, with parts that each has a value—such an attribute is called *complex*. Complex attributes have a schema of their own, too. For instance, in the people dataset, it could be the case that all records have a schema made up of attributes (name, address, age, weight, height) (it is customary to indicate schemas by listing attributes names within parentheses). An attribute like 'age' would have as values numbers like '16' and so on and thus would be simple. Conversely, an attribute like 'address' could be complex, with a schema like (street-number, street-address, city, zip code, state).

1.2.1 Structured Data

Structured data refers to datasets where all records share a common schema, i.e. they all have values for the same attributes (sometimes, such datasets are called *homogeneous*, to emphasize that all records have a similar structure).

Example: Structured Data

The following dataset is used in [20] (it comes originally from the US Bureau of Transportation Statistics, https://www.transtats.bts.gov/). It contains data about flights that departed from New York City in 2013; it is called `ny-flights` in this book. All the records have schema (flightid, year, month, day, dep_time, sched_dep_time, dep_delay, arr_time, sched_arr_time, arr_delay, carrier, flight, tailnum, origin, dest, air_time, distance, hour, minute, time_hour). The dataset has a dimensionality of 20 and a size/cardinality of 336,777 records. The first record is

```
(1, 2013, 1, 1, 517, 515, 2830, 819, 11, UA, 1545, "N14228",
"EWR", "IAH", 227, 1400, 5, 15, 2013-01-01 05:00:00)
```

The record is basically a list of 20 (atomic) values, one for each attribute in the schema. Thus, this is a structured, simple (tabular) dataset.

Example: Structured Complex Data

The following data record describes an imaginary employee; the schema is (first-name, last-name, age, address, department). Attribute 'address' is complex and has schema (street-number, street-name, city, state, zip code). Attribute 'department' is also complex and it has schema (name, manager). In turn, attribute 'manager' is complex and has schema (first-name, last-name).

```
("Mike", "Jones", 39, (1929, Main street, Anytown, KY, 40205),
("accounting", ("Jim", "Smith")))
```

Tabular data is structured data where all attributes are considered *simple*: their values are all labels or numbers without any parts. For example, the dataset `ny-flights` is considered tabular, while a dataset with records like the imaginary employee in the previous example would not be considered a tabular dataset, as some attributes are complex. As we will see later, this type of data is called *hierarchical*. It is sometimes possible to transform hierarchical data into tabular and vice versa.

When tabular data is in a file, each record is usually in a separate line, and inside each line, each attribute is separated by the next one by a character called a *delimiter*; usually, a comma or a tab. When using commas, the file is called a CSV file (for Comma Separated Values). If the schema itself (names of attributes) is included in the file with the data, it is usually in the first line—this is the reason it is called the *header*.

Tabular data is so-called because it is often presented as a table, organized into rows and columns. The rows correspond to data records and the columns to their attributes. Intuitively, each row describes an entity, or an event, or a fact that we wish to capture, with the attributes describing aspects of the entity (or event or fact). Not all data in tables follows this structure; data that does is called *tidy*, as we will see in Sect. 2.1.4.

Example: Data as Tables

The `ny-flights` dataset can be displayed as a table as shown in Fig. 1.2. It is customary to create a grid for rows and columns and to show the schema at the top, as the first row. In the figure we only show some of the schema (11 attributes) and 2 rows, for reasons of space.
Note that records (rows) fit neatly into the table because all records have the exact same schema. Also, values fit neatly in cells because they are all simple.

In most datasets, the size (recall: the number of records or records in the dataset) is much larger (at least one order of magnitude) than the dimensionality (recall: the number of attributes in the schema), but some datasets have high dimensionality

flightid	year	month	day	dep_time	sched_dep_time	dep_delay	carrier	flight	origin	dest
1	2013	1	1	517	515	2	"UA"	1545	"EWR"	"IAH"
2	2013	1	1	533	529	4	"UA"	1714	"LGA"	"IAH"

Fig. 1.2 NY-flights data as a table

(although what is 'high' depends on the context). As an example, the dataset **Chicago Schools**,[10] which contains school information for the city of Chicago during the 2016–2017 school year, has 661 rows and 91 columns (compare this with the `ny-flights` dataset). Analysis of datasets with a high ratio of attributes (columns) to data objects (rows) can be difficult, as we will see in Chap. 4.

Exercise 1.1 Get the Chicago Schools dataset and create a table for the first two rows. Just kidding! Pick your favorite dataset and create a table for two arbitrary rows.

1.2.2 Semistructured Data

Semistructured data is data where each record may have a different schema (also sometimes called *heterogeneous* data). In particular, some attributes may be *optional*, in that they are present in some records and not in others. Also, attributes may be a mix of simple and complex. Finally, some attributes may have as value *collections* of values as opposed to a single one (with each value in turn being simple or complex). As a consequence, records in semistructured data may have a complex structure.

In most real situations, semistructured data is used for datasets where attributes are not simple. Hence, it is very common to have datasets where the records, on top of being different from each other, have a complex structure.

Example: Semistructured Data

Assume a dataset describing emails. Each email has a *ID*, a *header*, and a *body*. The header, in turn, has attributes *timestamp*, *sender*, *receiver*, and *subject*. The timestamp attribute then has attributes *date* and *time*. Optionally, an attribute *CC* may be used; when used, this attribute may contain one value or a list of values. Such data is usually presented by showing both the attribute name and its value together, so as to avoid any confusion about what a value denotes. An example, with indentation used to display the structure and a colon (':') separating the attribute name from its value, would be[11]

[10] Available at https://bit.ly/30ZDJbf.

[11] This example is taken from the *Enron dataset*, a collection of emails from the Enron Corporation that has been used extensively by researchers for testing and analysis, as it is one of the few

```
ID: 19475126.1075855757890
Header:
        timestamp:
                date: Sun, Feb 4 2001
                time: 03:06:00
        sender: robert.benson@enron.com
        receiver: bsunsurf@aol.com
        subject: Rob-are you getting this?
        CC: peter.shipman@axiaenergy.com, gwadsworth@midf.com
Body: "How about lunch tomorrow?"
```

Note that attribute 'CC' has two values, separated by commas.

The above presentation mixes schema (attribute names) and data (values), while the tabular representation shows the schema once and does not repeat it. This is due to the fact that in tabular data, all records share the same schema, and therefore there is no need to repeat it for each record. On semistructured data, though, a record may be different from others; hence, we need to indicate, in each record, which attributes are present. Semistructured data is sometimes called *self-describing*, because each record contains both schema and data.

Semistructured data includes data in XML and JSON, two very popular data formats. They are very similar, differing only in how they present the data and a few other small details. XML describes the schema by using *tags*, labels that are enclosed in angular brackets. Tags always come in pairs, composed of an *opening* and a *closing* bracket. They can be identified because the closing bracket is exactly like the opening one but with the addition of a backslash. The value of the attribute goes between the tag pair.

Example: XML Data

The above email example is repeated here in XML format.

```
<email>
 <ID> 19475126.1075855757890 </ID>
 <Header>
        <timestamp>
                <date> Sun, Feb 4 2001 </date>
                <time>  03:06:00 </time>
        </timestamp>
        <sender>  robert.benson@enron.com </sender>
        <receiver> bsunsurf@aol.com  </receiver>
        <CC>
          <email> peter.shipman@axiaenergy.com </email>
          <email> gwadsworth@midf.com </email>
        </CC>
```

collections of real-life email datasets that is publicly available (see https://en.wikipedia.org/wiki/Enron_Corpus). Some fields are made up to simplify the example.

```
        <subject> Rob-are you getting this? </subject>
  </Header>
  <Body>  got it.        </Body>
 </email>
```

Note that indentation is no longer needed, as the tags clearly indicate the data for all attributes, simple or complex; it is used here only to aid legibility.

JSON uses a different format for the same idea. Instead of tags, JSON uses labels for attributes, which are separated by a colon (:) from their value. Also, when an attribute denotes a collection, the values are enclosed in square brackets ([]), and when they denote a complex object, they are enclosed in curly brackets ({ }).

Example: JSON Data

The XML data of our previous example can be written in JSON as follows:

```
 {
  ID: 19475126.1075855757890
  Header: {
         timestamp: {
          date: "Sun, Feb 4 2001",
          time: 03-06-00
         },
         sender:  robert.benson@enron.com,
         receiver: bsunsurf@aol.com,
         CC: [
           email: peter.shipman@axiaenergy.com,
           email: gwadsworth@midf.com
             ],
         subject: "Rob-are you getting this?"
          },
 Body:  "got it."
 }
```

Note how the whole thing is enclosed in curly brackets, as it denotes a single record. Note also how some values are enclosed in quotes, so that it is clear where they end (there are no ending tags in JSON). All values (even complex ones) end with a comma, with the exception of the last value in the array.

When complex attributes are present, the schema of a record can be described by what is called, in Computer Science, a *tree*: a hierarchical structure with a *root* or main part that is subdivided into parts. Each part can in turn be further subdivided.

Example: Tree Data

In Computer Science, it is customary to draw trees upside down: the root is at the top, and each level down a 'sub-part' of the one above. The parts at the very bottom, without any sub-parts, are called the *leaves* of the tree. In this example, the root is *Email* and the leaves are *ID, Date, Time, Address, Sender, Receiver, Body*. The node A above another node B is called the *parent* of B; B is the *child* of A. In a tree, each node has only one parent, but an arbitrary number of children. The parent of *Sender* is *Header*; the parent of *Header* is *Email*. The node *CC* has multiple children, indicated by the dots (...).

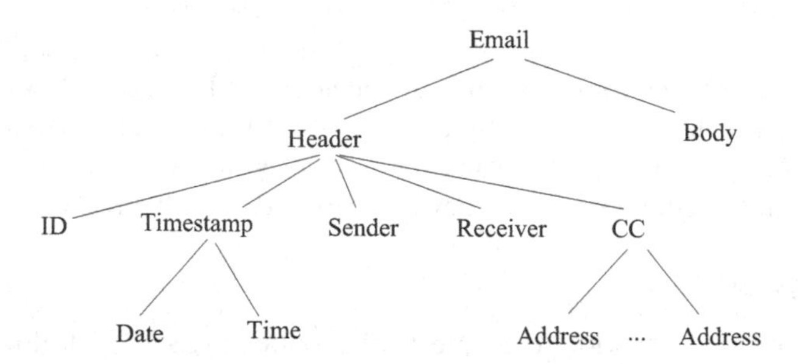

This tree represents the schema of the above examples; to add data, we would add a value to each leaf in the tree (since leaves represent simple attributes that have a simple value).

A semistructured dataset can be considered, then, as a collection of tree-like structures. It is important to note that in most practical cases the objects in the collection will have a good deal in common, that is, they all will share at least part of the schema. It is rare (albeit possible) to have an XML or JSON dataset where each object is completely different from all the others. This helps in dealing with semistructured data and makes it possible to put such data in a tabular format in many cases, as we will see in Sect. 2.3.1.

An important point to be made is that the border between structured and semistructured data is not rigid. It is possible to accommodate somewhat irregular data in tables, as we show next: give all records a schema with all attributes possible and leave empty cells for records that do not have values. Of course, this may yield a table with many empty cells, which makes analysis difficult, so this is only a good idea if data is more homogeneous than heterogeneous (items share many more attributes than not). As for attributes, whether an attribute is complex or simple is sometimes a design decision. For instance, an attribute 'address' could be expressed as simple (without parts) with values like "312 Main Street Anytown KY 40205" or the parts of it could be explicitly marked, as `street-number: 312, street-name: "Main Street", city: "Anytown", State: KY, zip: 40205`. As we will see, the choice to express the attribute in one way or another depends in part on what we want to do with the data. Even attributes whose value is a collection can be put in a tabular format, as we will discuss in the next chapter. In fact, one of the strengths of databases is that they make clear, in their design, what the data structure really is.

Example: Semistructured and Structured Data

The **Chicago Employees** dataset contains information about city employees in Chicago as of 2017, with a total of 33,693 records[12] The schema consists of attributes *Name, Job Title, Department, Full/Part Time, Salary/Hourly, Typical Hours, Annual Salary, and Hourly Rate*. Each record/row represents one employee. However, since an employee only has a salary if s/he is full time, and an hourly rate if s/he part time, and every employee is one or the other (an exclusive choice), not all employee records have values for all attributes. Two sample records are shown (names have been changed) in CSV format; attributes with no value are skipped (which creates the ',' followed by another ',' or by nothing if at the end of the line):

```
(Jones, Pool Motor Truck, Aviation, P, Hourly, 10,, $32.81)
(Smith, Aldermanic Aide, City Council, F, Salary,,$12,840,)
```

If we wanted to put this data in a table, we would have *missing values*. The same two records above are shown here; when there is a missing value in a cell, it is left empty.

Name	Job Title	Department	F/T	S/H	TH	AS	HR
Jones	Pool Motor Truck	Aviation	P	Hourly	10		$32.81
Smith	Aldermanic Aide	City Council	F	Salary		$12,840	

(we have shortened "Full/Part Time" to "F/T," "Salary/Hourly" to "S/H," "Typical Hours" to "TP," "Annual Salary" to "AS," and "Hourly Rate" to "HR"). This data can be put in XML or JSON format, since we can use the flexibility of XML/JSON to get rid of empty attributes by only listing, in each record, attributes (tags) for which values exist (also, if we assume that all part-timers are paid by the hour and all full-timers have a salary, we could get rid of an additional field).

```
<Employees>
 <Employee>
  <Name> Jones </Name>
  <JobTitle>  Pool Motor Truck    </JobTitle>
  <Department>  Aviation   </Department>
  <Full-Part Time> P   </Full-Part Time>
  <Salary-Hourly>  Hourly   </Salary-Hourly>
  <TypicalHours> 10  </TypicalHours>
  <Hourly Rate> 32.81   </HourlyRate>
 </Employee>
 <Employee>
  <Name> Smith </Name>
  <Job Title>  Aldermanic Aide    </JobTitle>
  <Department> City Council </Department>
  <Full-Part Time> F </Full-Part Time>
```

[12] Available at https://data.cityofchicago.org/Administration-Finance/Current-Employee-Names-Salaries-and-Position-Title/xzkq-xp2w.

```
    <Salary-Hourly>  Salary </Salary-Hourly>
    <Annual-Salary> 12,840 </Annual-Salary>
  </Employee>
</Employees>
```

Semistructured data can be used for tabular data, since semistructured data can accommodate cases where all records have the same schema, although in this case repeating the schema for each record is highly redundant. The idea is to treat the table as an object made of a repeated attribute; this attribute in turn is an object representing the row, with attributes for each column.

Exercise 1.2 Put the data about New York flights in XML and JSON.

One particular, important case of semistructured data is *graph* data. Graph data (sometimes called *network data*) represents collections of objects that are connected (or linked, or related) to each other. A fundamental part of the dataset, then, is not just the objects themselves, but their connections. In many common situations, each connection links a pair of objects; the graph can be seen as a collection of objects and their pair-wise connections.[13] In the context of graph data, the objects are called *nodes* or *vertices*, and the links are called *edges*.

Example: Graph Data

The typical example of graph data nowadays is a *social network*, where nodes represent people and edges represent relationships between two people. In a scenario like Facebook, for instance, each node represents a Facebook user, and two nodes A and B may have an edge between them ('be connected') if A has declared B to be a friend, or if A has liked one of B's posts. In Twitter, nodes also represent users; two nodes there can be connected if A has re-tweeted or liked something that B tweeted, or A 'follows' B, or A and B share a hashtag.

Assume Shaggy, Fred, Daphne, and Velma are Twitter users, related as follows (we give a list of edges, with labels to indicate the type of relationship):
(Shaggy, Fred, follows)
(Shaggy, Daphne, follows)
(Shaggy, Velma, follows)
(Daphne, Velma, re-tweets)
(Fred, Daphne, likes)

This is usually depicted in a diagram, typically using circles for the nodes and arcs between them for the edges.

[13]It is possible for a connection to link more than two objects; the typical example is a relation SUPPLIES that connects suppliers, parts, and projects (3 objects). Graphs are limited to binary (two objects) relations, but are still very useful.

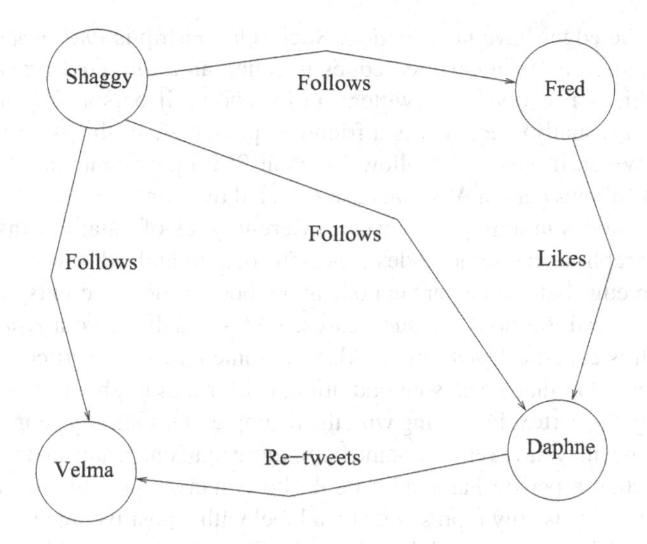

In many datasets, a node may have (regular) attributes to represent information about the object it stands for. In the case of Facebook data, for instance, each node may contain the information each user added to her *About* section, including work and education data. What makes this a graph, though, is the links to other users.[14]

One way to think of graph data is as records where the value(s) of some attributes are other records. For instance, in the previous example we can think of each node as a 'person' record where some attributes (like "follows") have as the value another person. Seen this way, graph data is semistructured because a node may have any number of such attributes—since a node may have an arbitrary number of edges to other nodes.

What makes graph data special is that the relationships between nodes are considered particularly important, so most analysis focuses on properties related to the edges, like finding whether two nodes N1 and N2 are *connected* (i.e. whether there is a sequence of edges, called a *path*, leading from N1 to another node N,' and from N' to another node N,"..., and so on until one last edge leads to N2), or how many connections there are from a given node to others. We discuss graph processing in Sect. 4.6.

A few variations of this idea are useful. Sometimes the edges between nodes have a *direction*, that is, they represent a one-way relationship. Such relationship is *directed*; a graph made up (exclusively) of directed edges is called a *directed graph*.

[14]This example is a simplification. Both Facebook and Twitter keep a very detailed dataset for each user, including every photo, video, or text that the user has ever posted, all their profile information, and everyone you have ever declared a friend (plus, in the case of Facebook, information about advertising categories that Facebook thinks you fit).

Other times, the edges have no direction. Such relationship is *undirected*; a graph made up (exclusively) of undirected edges is called an *undirected graph*. A good example of this is Facebook vs. Twitter. On Facebook, if person A is a friend of person B, then (usually) person B is a friend of person A, so this is an undirected relation. In Twitter, if person A 'follows' person B, it may or may not be the case that person B follows person A, so this is a directed relation.

Sometimes nodes in a graph represent different types of data. For instance, we may have a graph where some nodes represent people and other nodes represent books, and an edge between a person node and a book node represents the fact that the person has read the book. In such cases, nodes usually have a *type* attribute, and the graph is called a *typed graph*. Also, in some cases the vertices of a graph have *labels*, a value that gives some additional information about the connection represented by the vertex. Following with the example of books and people, suppose that some people may have read the same book more than once, and so it is necessary to indicate, when a person has read a book, how many times this has happened. Such information is usually represented by a label with a positive integer value. The graphs that use labels are called *labeled graphs*. The graph depicted in the previous example is directed and labeled.

It is interesting to note that, technically speaking, trees are special types of graphs. In a tree, edges are directed in the direction parent $\rightarrow$ son, and each edge must have only one parent (although it can have an arbitrary number of sons). Also, trees do not admit *cycles* (paths that end up in the same node that they started). This implies that no node can appear as its own *descendant* (a son, or a son of a son, or a son of a son of a son, etc.) or as its own *ancestor* (a parent, or a parent of a parent, or a parent of a parent of a parent, etc.). Hence, all information coming in XML and JSON can be considered 'graph data.' However, this is not how such data is treated. For one, trees are much simpler to deal with than arbitrary graphs, so it is important to note when data schema forms a tree and when it forms a graph. The reverse of this is that graph data can handle some things that are difficult to represent as a tree, as we will see in the next chapter.

1.2.3 Unstructured Data

Unstructured data refers to data where the structure is not *explicit*, as it is in structured and semistructured data (that is, no tags or markers to separate the records into attributes). Most of the time, unstructured data refers to *text*, that is, to sentences written in some natural language. Clearly, such data has structure (the grammar of the language), but the structure itself is not shown. Thus, in order to process such data, extra effort is needed.

Text data tends to come in one of the two ways: First, when we have a collection of documents (usually called a *corpus*), the text in the documents themselves is the main target of analysis. Second, we may have tabular data where one or more of the attributes are textual in nature, as in the email dataset example, where the *subject* and *body* attributes can be considered text, as in the next example.

Example: Text Data

The **Hate Crime** dataset, from ProPublica,[15] consists of 2663 records, each one describing an article in a newspaper. All records have the same schema, composed of attributes *Article-Date, Article-Title, Organization, City, State, URL, Keywords, and Summary*. Each attribute is simple because its value is a name or something similar. The first record, with each value in a separated line (and the last two enclosed in double quotes, for legibility), is

```
  3/24/17 13:10,
Kentucky Becomes Second State to Add Police to Hate Crimes Law,
Reason,
Washington,
District of Columbia,
http://reason.com/blog/2017/03/24/kentucky-becomes-second-state
-to-add-pol,
“add black blue ciaramella crime delatoba donald gay hate law
laws lives louisiana matter police trump add black blue
ciaramella crime delatoba donald gay hate law laws lives
louisiana matter police trump”,
"Technically, this is supposed to mean that if somebody
intentionally targets a person for a crime because they're
police officers, he or she may face enhanced sentences for a
conviction. That's how hate crime laws are used in cases when
a criminal targets"
```

Note that the second attribute, *Article-Title*, and the last two attributes (*keywords* and *Summary*) can be considered text (*Keywords* is, as its name indicates, a list of keywords, and *summary* is a brief description of article content that is two sentences long).

We discuss how to store text in Sect. 2.3.3 and how to analyze it in Sect. 4.5.

Again, it is important to remember that the same dataset can be structured in different ways. For instance, even though Twitter data can be seen as a graph, Twitter actually shares its data using JSON: each Tweet is described as a JSON structure with fields like *text, user* (with subobjects `id`, `name`, `screen_name`, `location`), *entities* (with subobjects `hashtags` and *urls*, both of them collections). As another example, the `ny-flights` data introduced in a tabular way in Example 1.2.1 could also be presented as graph data by making the airports the nodes and

[15] This dataset can be obtained at https://www.propublica.org/datastore/dataset/documenting-hate-news-index.

considering each flight an edge from airport A to airport B, using the information in attributes `origin` and `dest`. Note that this is a directed graph, and that the edges do not have just a single label, but a collection of them (all other attributes). This format could be more appropriate than the tabular one for certain types of analysis. This lesson carries one very important consequence: when we acquire data, it will come in a certain format. This format may or may not be the appropriate one for the analysis that we want to carry out. Hence, it will sometimes become necessary to transform or *restructure* the data from one format to another. We will discuss this task in Sect. 3.4.1.

Exercise 1.3 A description of Twitter data in JSON can be found at
https://developer.twitter.com/en/docs/tweets/data-dictionary/overview/user-object.
Using that description, build a small JSON dataset with two imaginary users.

Exercise 1.4 Take a few rows of the `ny-flights` data from Example 1.2.1 and create a diagram to show it in graph format.

One final observation is that some datasets come as a mix of formats, which usually renders them useless for analysis. Typical examples include data from *spreadsheets*, very popular tools for dealing with data. Even though many datasets look like a collection of columns organized by rows, spreadsheets are structured differently—in fact, they are not structured at all. A user is free to put whatever she wants in any cell of the spreadsheet; as a result, columns many not contain the same type of data at all. Also, rows are not required to look like other rows. Spreadsheets and their problems are discussed in depth in Sect. 2.1.4.

Example: Spreadsheet Data

An example of data in spreadsheets is the ARCOS (Automation of Reports and Consolidated Orders System) report from the U.S. DEA (Drug Enforcement Agency). It is organized as a set of pdf documents, one per year. In each document, the report is broken down by drugs, and within drugs, it provides data per state (broken down by zip code and quarter of the year). A sample page is shown below. Such data is nearly impossible to use as provided, due to the format (we need the data in alphanumeric format, while the pdf is more like multimedia, in that it is encoded using a proprietary code) and to the lack of structure (even though the data *looks* tabular, it really is not, presenting issues similar to those of spreadsheets).

DATE RANGE: 01/01/2017 TO 12/31/2017 | ARCOS 3 - REPORT 1 | Run Date: 07/03/2018

RETAIL DRUG DISTRIBUTION BY ZIP CODE WITHIN STATE BY GRAMS WT

DRUG CODE: 1100 DRUG NAME: AMPHETAMINE

STATE: ALABAMA

ZIP CODE	QUARTER 1	QUARTER 2	QUARTER 3	QUARTER 4	TOTAL GRAMS
350	17,317.77	17,592.62	17,601.30	18,026.63	70,538.32
351	8,853.04	8,730.36	8,871.69	8,885.75	35,340.84
352	19,011.67	18,775.54	19,012	19,869.23	76,668.44
354	8,415.59	8,008.47	7,903.15	8,576.03	32,903.24
355	3,437.61	3,592.54	3,505.06	3,423.91	13,959.12
356	9,419.78	9,504.50	9,896.92	10,055.83	38,877.03
357	4,092.23	4,275.89	4,362.04	4,447.32	17,177.48
358	5,351.36	5,555.98	5,650.23	5,681.07	22,238.64
359	4,819.79	4,836.55	4,619.88	5,088.01	19,364.23
360	6,516.37	6,441.69	6,372.90	6,768.46	26,099.42
361	5,920.66	5,825.58	5,225.22	5,243.17	22,214.63
362	2,442.19	2,350.13	2,347.92	2,345.96	9,486.20
363	5,219.18	5,291.86	5,210.23	5,424.80	21,146.07
364	1,273.24	1,315.80	1,144.04	1,299.77	5,032.85
365	11,488.92	11,634.86	12,049.97	12,420.28	47,594.03
366	7,815.82	7,687.68	7,886.41	7,953.48	31,343.39
367	1,557.50	1,463.38	1,478.39	1,572.70	6,071.97
368	6,415.23	6,190.82	6,134.74	6,677.12	25,417.91
369	154.58	123.62	127.6	143.08	548.88
TOTAL	129,522.53	129,197.87	129,399.69	133,902.60	522,022.69

STATE: ALASKA

ZIP CODE	QUARTER 1	QUARTER 2	QUARTER 3	QUARTER 4	TOTAL GRAMS
995	4,873.14	4,889.14	4,867.94	5,002.53	19,632.75

1.3 Types of Domains

As we have seen, structured and semistructured data are made up of basic building blocks: individual attributes that represent features or characteristics of data records. Complex attributes are made up of schemas of simple ones; in the end, each simple attribute represents a *domain* or set of values. For a given data record, the attribute provides a value from its domain. For instance, in the tabular dataset with people information, the attribute 'height' has values that provide the height of a particular person, expressed by numbers like 5.8 (if we assume that heights are measured in feet and inches) or 183 (if we assume that heights are expressed in centimeters). We would expect all such values to be numbers and to have a certain range. Thus, we can say that the attribute represents a *numerical* domain, expressed by real, positive numbers with one digit precision and up to a certain value. Conversely, an attribute like 'name' would have values like "Smith"; this domain is not numerical, and in principle it seems impossible to determine what values are in it (even something like "Xyz" could be a name in some languages). It would make sense to compare two heights and see which number is larger. It would not make sense to compare two names that way. It is clear that domains are of different types, and that different operations can be applied to different domains. Hence, recognizing different domains is an essential part of data analysis. Roughly speaking, domains for simple attributes fall into one of the three categories: *nominal* (also called *categorical*), *ordinal*, and *numerical*. Some people call nominal data qualitative and numerical data quantitative.[16] We explain each domain in more detail next.

1.3.1 Nominal/Categorical Data

Nominal/categorical data refers to sets of labels or names (hence the 'nominal'): the domain is a finite set, with no relations among the elements. In particular, we cannot assume that there is any order (a first element, a second, etc.) or that elements can be combined. A typical example is an attribute 'Country' defined over the domain of all countries.[17] Note that we can put country names in alphabetical order; however, this order does not correspond to anything meaningful in the domain (it does not organize countries in any significant way). Another typical example is an attribute within a dataset of emails that gives the type of each email as one of 'work,' 'personal,' 'spam.' Here, the labels represent *categories* or classes of email (hence the 'categorical'). Again, we could order these labels alphabetically, but this order really carries no meaning. The most representative characteristic of nominal/categorical attributes is precisely that we cannot do anything with them except ask if a given label belongs to the domain or if it is present in the dataset, and whether the labels of two records are the same or not. We can also count how many times a given label is present in a dataset and associate a *frequency* with each label. Beyond that, very few operations are meaningful: finding the mode or calculating the entropy of the associated frequencies or applying the χ^2 test are the most common.[18]

Two things are worth remarking about this type of domain. First, values of a nominal/categorical attribute may be represented by numbers, but that does not make them numerical attributes. This is, in fact, quite common; it is called a *code*—for instance, '0' for 'married' and '1' for 'not married.' Again, even though they look like numbers, they do not behave like numbers. As another example, a zip code may look like a number (a string of digits), but it is not. To see the difference, ask yourself if it makes sense to add two zip codes or multiply them—operations that you would normally apply to numbers. Another characteristic worth mentioning is that sometimes there may be a hierarchical organization superposed on a nominal/categorical attribute. Consider, for example, a 'time' attribute that represents the months of the year. They can be organized into seasons (spring, summer, fall, winter) or quarters. Or a *location* attribute, where a point can be

---

[16] There is no total agreement on the details of this division. For some people, qualitative data includes both nominal and ordinal data.

[17] Underlying domains may be hard to define. For instance, in this example one may ask whether this refers to *current* countries, or any country that has ever existed. For more on this issue, keep on reading.

[18] Here and in the next few paragraphs we mention several statistical operations; for readers not familiar with them, these are all discussed in the next chapter.

located within a city, the city within a county, the county within a state, the state within a country. When looking at such data, it is possible to look at the dataset from several levels of the hierarchy; this is usually called the *granularity level*. Moving across levels is, in fact, a common analysis technique.

1.3.2 Ordinal Data

Ordinal attributes are, like nominal ones, sets of labels. However, a linear ordering is available for the domain. We can think of a linear order as arranging the elements in a line (hence the "linear"), with elements 'going before' others. Technically, a linear order is one that is irreflexive (no element goes before itself), asymmetric (if A goes before B, it cannot be that B goes before A), and transitive (if A goes before B and B goes before C, then A goes before C). As a consequence of having an order, we can always enumerate all elements in the domain and distinguish between a first one, a second one, ... and a last one.[19] The order also allows us to compare values with respect to their position in the order. Examples would be classifying symptoms of a sickness as very mild, mild, medium, severe, very severe; or opinions on a subject as strongly agree, agree, neutral, disagree, strongly disagree. Note that, since values can be ranked, we can compute a *median* (and, in the associated frequencies, a mode) and also percentiles and rank correlation, but we cannot do further operations. This type of attribute also includes codes; for instance, using '1,2,3,4,5' for 'very satisfied,' 'satisfied,' etc. These codes are still not numbers.

Ordinal attributes occupy a gray space between nominal and numerical attributes. They are labels, but the fact that they can be ordered means that we can, to a small degree, treat them as numbers. In fact, numbers are also ordinal attributes, as they can always be put in the usual linear order < ('less than').

1.3.3 Numerical Data

Numerical attributes are representations of quantities, dimensions, etc. and are truly represented by numbers, be they integers or reals.[20] It is customary to further distinguish two types of numerical attributes:[21]

- interval: in these, the zero value (i.e. a point where all values start) is arbitrary. As a consequence, the distance between one value and another is (approximately) the same, so we can compare values and add/subtract them, but taking ratios

[19] At least for finite domains, which are the only ones of interest here.

[20] Note that reals can only be represented in a computer with a certain degree of precision. This will be important for numerical computations, and we will discuss it later.

[21] Some statistics textbooks distinguish between interval, ratio, and absolute types.

make no sense. We can take the mean, median, and value, as well as range and standard deviation/variance. Hence, we can apply the t-test and calculate Pearson's correlation coefficient. A typical example is a 'date' attribute: there is no 'zero' date, but the difference between the two given dates can always be calculated. Another example is the temperature measured in Celsius or Fahrenheit: while both scales have a zero degree mark, they both continue below zero: zero is not the absolute lowest value, and the place where the zero is somewhat arbitrary (as shown by the fact that the 'zero' in Celsius and the 'zero' in Fahrenheit are different).

- ratio: in these, there is an absolute zero, hence ratios make sense. A typical example is the temperature measured in Kelvin degrees (in this scale, zero marks the point where there is no thermal motion, and therefore it is impossible to go below it). In fact, most measurements (of time, distance, mass, money, etc.) fit here. For instance, a time of 0 s means no time has passed; a time of 10 s is twice as much as 5 s. With this domain, we can compare the values, add/subtract, multiply/divide, find mode, median, and mean (including arithmetic and geometric mean), as well as pretty much any mathematical manipulation. To note, the counts that result from associating frequencies to categorical values are in this category, which is why calculating frequencies is so useful and so commonly done.

Determining the type of a domain is extremely useful for analysis since it gives us an idea of what operations make sense for an attribute.

Example: Domains in Datasets

In the `ny-flights` dataset, all attributes are atomic. Each one represents a domain, but most attributes are temporal (`year, month, day, dep_time, sched_dep_time`, and so on). As already stated, dates and times are all numerical (interval type, to be precise), but by itself we can consider `month` a categorical attribute.

In the Chicago employees dataset, attributes `Name, Job Title, Department, Full-Part Time`, and `Salary-Hours` are categorical (attributes `Full-Part Time` and `Salary-Hours` can only have two values each; these are sometimes called *binary*). The rest of the attributes (`Typical Hours, Annual Salary, Hourly Rate`) are numerical—in fact, all of them are ratios, as their zeros are truly the 'zero' of each domain. The combination of categorical and numerical domains is extremely common in datasets for analysis.

Exercise 1.5 Classify the domains of all the attributes in the schema of `ny-flights`.

Exercise 1.6 A car database has an attribute called "Maintenance." Typical values are "60000 km/2 years," "80000 km/3 years." Is this a categorical or numerical attribute? What should it be for analysis purposes?

For practical purposes, it is a good idea to examine the domain from the point of view of the possible values allowed—equivalently, how membership in the domain is determined. Roughly, we can distinguish between:

- *closed* sets: also called *enumeration* domains. These are domains where the set of labels allowed can be precisely given. An example are the cities in a country or the months of the year. These domains tend to be static, i.e. do not change over time. There is usually some external resource that can help determine membership, which is simply a matter of looking the label up in the resource. Even though these simple sets can present issues, determining membership is usually not hard. For instance, the list of cities in a country may be different under different names—a list of German cities in German vs one in English, but once the language has been fixed, we can determine valid labels.
- *bounded* sets: these are domains where the elements are taken from a larger domain but somehow limited in some way. A typical example is *age*, which is expressed by a number—but not any number. For instance, the age of a person, expressed in years, can be any number between 0 and, say, 120. Negative numbers are excluded since they do not make sense; fractions and reals are also excluded due to the convention of expressing age in rounded amounts. Very large numbers (like 5,000) are out of bounds (this would be a typical example of an outlier). The difficulty of determining membership in these domains depends on whether the bounds are fixed or variable, strict or fuzzy.
- *patterned* sets: these are domains where all elements must obey a certain pattern (or one of a finite number of patterns). Typical examples are email addresses and phone numbers. The difficulty of determining membership here depends on the complexity of the patterns. For instance, for regular mail addresses in one country, the complexity tends to be low; however, mail addresses over the world follow a bewildering variety of patterns, and determining whether an international address is correct can be extremely hard.
- *open* sets: these are sets where it is not possible to give a definite criterion for membership. A typical example is person names. What counts as a person's name, even in a given language, is subjected to change over time (i.e. these sets tend to be dynamic) and is open-ended (in some countries, parents can give their children whatever they want as a name, i.e. they do not have to follow any rules). Consequently, determining membership may be very difficult or even impossible.

The importance of these distinctions is that one of the main tasks during EDA and cleaning is to make sure that all values in our data are correct for their domain. As we can see, deciding this depends on the type of the underlying domain, and it may be extremely simple or plain impossible.

Example: Simple Domains

In the Chicago employees dataset, attribute `Name` is an open set, while `Job Title`, `Department` are likely closed (that is, there is a finite list of possible job titles and possible departments). `Full-Part Time` and `Salary-Hours` are also close;

`Typical Hours`, `Annual Salary`, `Hourly Rate` are very likely bounded, in that we can determine minimum and maximum values for each. If this is correct, we can determine whether values in our data are within permissible bounds or they are the result of some error.

Exercise 1.7 Classify the domains of the attributes in the schema of `ny-flights` from this point of view.

1.4 Metadata

Metadata is *data about data.* It describes what the data refers to and its characteristics as a whole. Having metadata is highly beneficial for several reasons:

- To operate meaningfully in data, we must only carry out operations that make sense for that data: it makes no sense to add a temperature and a height, even though they are both numbers.
- In general, having metadata can guide our decision as to how to clean and preprocess the data.
- Sometimes data analysis will reveal that a transformation was not quite appropriate; we may want to *undo* the effects of the change and perhaps try something different. This is much easier to do if we recorded clearly what the changes were in the first place.
- If we need to share/export our data, we need to explain what the data are (what they mean or refer to) for others to be able to use them. Data storage and manipulation always happens in the context of a project that has certain goals. However, very frequently data collected for a certain purpose is reused later for different projects, with different goals. Hence, it is important to understand what the data was originally meant to represent, and how it has been transformed, in order to assess suitability for different purposes than the original one.[22]
- Because it provides a trace of how data came to be, metadata is crucial in helping with *repeatability* of analysis in order to generate *reproducible* results. This is becoming more and more important in all kinds of analysis.

In spite of its importance, metadata has traditionally been ignored in data projects. One reason is that it is typically considered overhead, i.e. work that gets in the way of obtaining results from the data. Another significant reason is that there is very little agreement among experts as to what constitutes good metadata, and how it should be generated and stored. However, given its potential positive impact, an attempt should be made to manage the metadata of any given dataset. In this section,

---

[22]This is a field of study on its own right, called *Data Reuse*.

we give some basic guidance as to what metadata should be kept and when it should be created; later in Sect. 3.5 we discuss how to store and manage it.

The concept of metadata is vague and it can be extended to cover many aspects of data and data processing. As a consequence, there are many diverse classifications of metadata in the literature. However, there are some basic parts that enjoy wide support:

- *structural metadata*: for both structured and semistructured data, metadata should include the schema or a given dataset[23] and also a description and classification of the domain of each attribute in the schema.
- *domain metadata*: Domain information is especially useful. Two aspects merit mention: *syntactic* and *semantic*. Syntactic metadata refers to how data is represented, what kind of values it can take. For numeric values, this includes appropriate range and typical values; in the case of measures, *unit*, *precision*, and *scale* are a must. For instance, a field giving salaries may have values in Canadian dollars, representing thousands (so that the value '85' actually means 'CAN $85,000'); a field giving people's heights (an example we have used before) should, at the minimum, describe which units are involved: metric ones (meters and centimeters) or English units (feet and inches).[24] For categorical values, this may involve (depending on the kind of domain) a list of valid values or valid patterns; this will allow us to check for correct data. For instance, a *name* field may come with a description of what a good value is supposed to look like: last name, followed by a comma, followed by a first name, with both names capitalized. For dates, a description of the format (for instance, 'month-day-year') is highly desirable, as this avoids (typical but painful) confusions.

  Semantic metadata refers to what the data is supposed to represent. For instance, knowing that an attribute is for people's names or is a measurement of people's height helps considerably when examining the data since in many cases the analyst can think of typical values, values that may not be correct, etc. Note that in such cases the name of the attribute ('name,' 'height') tends to be enough to point us in the right direction, but even when using meaningful names, this may not be enough. For instance, an attribute named 'price' may give the price of a product, but this may be before or after taxes, with or without discounts.[25] In the case of codes, semantic metadata is especially useful, as many codes tend to have cryptic names. For instance, in a list of products, a numerical attribute called *FY15* may be an enigma, until we find out that the name refers to 'Females Younger than 15' (and not, say, to 'Fiscal Year 2015'). The problem with codes may be in the values themselves; for instance, an attribute *Customer Satisfaction* may have values 1–5, which need to be interpreted (while this is a typical ordinal

[23] In the case of semistructured data, each record may have a different schema, but in most practical cases most records share a common schema with a few variations.

[24] Note that in cases where we have a good knowledge of the domain it may be easy to deduce unit and scale by looking at some data; however, this will not always be the case.

[25] In fact, a product may have many prices in a given dataset, because of these distinctions.

domain, it is important to know if bigger number means more satisfaction or less satisfaction). Thus, explaining what the data is supposed to represent in plain and clear language is an invaluable aid for any kind of analysis.

- *Provenance/lineage metadata*: as stated above, this describes where data comes from. Provenance refers to two different, but related aspects: for raw datasets, it refers to the source of the data; for clean/processed datasets, to the manipulations that resulted in the current state of the data. In the first sense, provenance is a way to explain how data was obtained. It should describe the source (for instance, a sensor, a web page, or a file), the date when data was obtained, and other relevant information (for instance, the original owner and/or creator of the data).[26] It is especially useful to list constraints and assumptions under which the data was generated. For instance, data from an ER (Emergency Room) in a hospital should identify the hospital, the schedule under which data was collected, and whether the original source was original medical transcripts, electronic records, etc. The second sense of provenance applies to datasets generated by manipulating data through the data life cycle process: by cleaning, pre-processing, analyzing, etc. This includes recording actions applied to the data and their parameters. For instance, an attribute may have been found to include many abnormal values; a decision is made to delete them and substitute them by the mean (average) of the remaining values. Whether this is a good decision or not depends entirely on the context. Thus, the best thing to do is to document this step; if later on it turns out to be misguided, we may be able to undo it and try something else.[27]
- *Quality metadata*: this is a description of how good the data is. This metadata may not be available at all when data is acquired and may have to be generated after the EDA step. Nevertheless, it is an important aspect to document, since many datasets are found to contain problems that affect their usefulness. In fact, this is such a common issue that a whole sub-field, *Data Quality*, has developed concepts and techniques that deal with problems in the data. While quality is a many-faceted concept, some key components have been identified by researchers [2] and should be present in the metadata, if at all possible:
  - Accuracy: how close the represented value is to the actual value. This is especially important for measurements and may require specifying how values were acquired. In the case of continuous domains, there is usually an imposed limit to the accuracy; this should be documented. Accuracy will help determine the *precision* of the values.
  - Completeness: this refers to "the extent to which data are of sufficient breadth, depth, and scope for the task at hand" [2]. In a relational database, we can talk of *schema completeness* (do we have all entities and attributes needed?), *column completeness* (how many values are missing in an attribute?), and

[26]Such information is very useful to, for instance, have someone to go and ask questions.

[27]Note that recording an action is no guarantee that it can be undone (some actions cannot be reversed); however, not recording it pretty much guarantees that the action cannot be undone and, worse, that trying to identify and mitigate its consequences will be almost impossible.

population completeness (do we have all values from the domain?). The latter is especially important since many datasets are *samples* from an underlying population, and often they have been obtained in a process that mixes opportunistic and random characteristics.

- Consistency: this measures whether the data as a whole is sound; it answers the question: are there contradictions in or across records? Sometimes a record may contain inconsistencies; for example, a person record with a *name* like "Jim Jones" and "female" for *sex*; or a record with '5' for *age* and 'married' for *marital-status*; or an order record where the date when the order has to be delivered is earlier than the date the order was placed. Sometimes two or more records may contradict each other (in a dataset of flight information, there may be two records sharing the same identifier but different destinations). Inconsistencies in the dataset need to be eliminated before analyzing the data, as they negatively affect pretty much any type of analysis.
- Timeliness: this refers to value changes along time. We want to know, mainly, two things: how often data changes (*volatility*) and how long ago was data created and/or acquired (*currency*); the two of them together determine whether the data is still valid (and, in general, how long it will remain valid). Volatility refers to the frequency with which data changes (rate of change). Data can be *stable* (volatility: 0), *long-term* (volatility: low), *changing* (volatility: high). Currency refers to when data was created, but may also refer sometimes to when data was acquired for the dataset. For instance, assume a sensor that takes temperature measurements every hour. Usually, a *timestamp* attribute will tell us when the values of the *temperature* attribute were taken. Assume, further, that the sensor sends information wirelessly every 24 h to a database. Thus, the acquisition time is the same for a whole set of data. Note that data may be current and not timely (for instance, a schedule of classes is posted after the semester starts); this is because timeliness depends on both currency *and* volatility. Note also that if data is updated, this affects its currency.
- Certainty: how reliable is the data (how much do we trust the source)? For many measurements, this is not an issue, but in certain analysis it can become a crucial element. For instance, when analyzing news reports, we may find out that reports on certain subjects (for instant, sudden, high-impact events) are highly uncertain.

It is clear that these aspects are related: if the data is uncertain, we cannot estimate its completeness, but if it is inconsistent, it cannot be certain. Also, accurate data tends to be certain; it is unlikely (although possible) that we have a very accurate measurement but are uncertain about its reliability. Thus, it may be tricky to evaluate all the aspects in isolation. Also, not all aspects apply to all datasets. Thus, one may want to create only essential metadata for a given project. However, all shortcuts taken will restrict our ability to reuse and publish the data.

Some authors will consider, beyond what we have outlined here, other aspects like *administrative metadata*—a description of rights/licenses, ownership, permissions, regulations that affect the data, etc. This may be advisable for data subjected to legal or ethical rules, for instance, private data that must be kept out of reach of the general public.

Exercise 1.8 Generate domain metadata to describe, as best as you can, the meaning of the attributes in the schema of `ny-flights`.

When is metadata generated? Ideally, metadata should be created at the acquisition/storage stage, when data is first acquired. This metadata should reflect what we know about the source of the data and the domains that the data is supposed to represent, that is, the *domain knowledge*. Having metadata that describes the data is very useful for data exploration, cleaning, and pre-processing. In particular, dealing with missing values and outliers is much easier when we have an idea of what data is supposed to be like. This initial metadata can be refined by EDA. For instance, suppose that we are gathering physiological data for a medical study, and one of the features we have is blood pressure. This is measured with a couple of numbers that give the *systolic* and *diastolic* pressure, and for most people is around 120 (for systolic) and 70 (for diastolic). Numbers up to 130 and 85 are still considered normal, but numbers above those indicate hypertension (high blood pressure). Conversely, numbers below 120 and 70 may indicate hypotension (low blood pressure). Knowing all this can help us determine if some values are outliers (for instance, numbers like 300 or 20 are probably the result of some error) or extreme values (for example, numbers like 180 and 110 indicate extreme hypertension, but are certainly possible). Note that this information comes from the domain knowledge and needs to be reconciled with what EDA extracts from the data. For instance, if our dataset is from people with heart disease, it could be the case that most of them have high blood pressure. Or, if the dataset is about infants or school-age children, the data will have different values. If EDA finds out that the metadata is not a good description of the data, we must reconcile the observed measurements with the assumed interpretation of the data.

During cleaning and pre-processing we may apply several changes to the data (see Sects. 3.3 and 3.4 for details). These changes should also be incorporated into the metadata, as already discussed.

As analysis proceeds, the methods used, together with any parameters and assumptions, should also be recorded. This will contribute to a correct interpretation of the results.

When it comes to archiving, metadata itself should always be a prime candidate to be kept. Having provenance metadata may allow us or others to recreate the original dataset (if not kept) as well as all the work done. Metadata is also the main tool to determine whether data can be reused for purposes different from the ones that motivated its collection—in particular, structural and domain metadata will ensure that the dataset is correctly understood. Finally, metadata tends to be much smaller than the data itself, and therefore it can be easier to store.

1.5 The Role of Databases in the Cycle

A database can be used in several roles in Data Analysis, depending on the exact situation. The first scenario is when the data already exists in a database; therefore, we have to go there to access it. After accessing the data, we have two options: one is to export the data to files (see Sect. 2.4 for details) and use R or Python or some other software to do all the work (see Chap. 6 for a brief description of how R and Python can interact with a database). Another option is to carry out some of the work (for instance, some EDA, some data cleaning and pre-processing) in the database and then export only the clean, relevant data to a file and continue with other tools. When the analysis does not include sophisticated Machine Learning or Data Mining techniques, we may be able to do all the work inside the database, avoiding the extra effort to export the data and use a different tool (see Chap. 4). Even when this is not the case, doing some of the work within the database still offers a number of advantages over tools like R and Python. First, when the dataset is very big, some preliminary work may allow us to extract only a part of the data (either a sample or a carefully chosen subset), which would be beneficial for further work, as neither R nor Python is particularly suited to work with large datasets. Second, databases offer strong *control access*: we can carefully monitor and limit access to the data, an advantage if there are concerns about data confidentiality. Third, if the data arrives periodically, or all at once but is later updated (that is, if data changes at all), the database offers tools to handle this *evolution* of data that do not exist in R or Python. Certainly, changes can also be managed at the file level, especially if one is knowledgeable about the power of the command line [9], but it is certainly nice to have tools that make life easier (see Sect. 2.4.2 for information about modifying existing data).

In a second scenario, data is to be captured yet and it is decided to use a database because of the size of the dataset or its complexity. In that case, we start by moving the data into the database (again, see Sect. 2.4 for details). Once this is done, we are in the previous scenario, and the same considerations apply.

In many cases right now data exists in files and the whole process takes place in R or Python. That is, databases are absent from the process. While this is totally justifiable in many cases (small datasets, complex, or ad hoc processing), this approach is manual-intensive and tends to produce results that are not well documented and are difficult to reproduce unless the markup tools available in R and Python are used.

All in all, databases can be a helpful tool in managing data through its life cycle, many times in combination with other tools. The next chapter goes through the process of putting data in the database, updating it if needed and exporting it. Later chapters discuss how to examine, clean, and transform the data for analysis.

1.5 The Role of Databases in the Cycle

[illegible]

Chapter 2
Relational Data

Relational database systems store data in repositories called *databases* or *schemas*. The latter name is due to the fact that (as we will see) data repositories have schemas, like datasets. In this book, we use 'database' for the data repository and reserve 'schema' for the description of the structure.

After starting a database server and connecting to it (see Appendix A for instructions), it is necessary to create a database, which is done with the SQL command

```
CREATE DATABASE name
```

or, alternatively

```
CREATE SCHEMA name
```

The name is just a label, but it is customary to give databases descriptive names that bring to mind the data they contain.

It is possible to create several databases in the same server. In that case, it is required that each database be given a different name. Whenever a user connects to a database server, she must specify which database to work on by giving the name: in Postgres, the command is

```
connect database-name
```

and in Mysql,

```
use database-name
```

After that, all commands that are explained in this chapter and the next one will work inside that database (it is also possible to work with several databases by switching from one database to another).

Once a database is created, the next step is to create *tables*. This can be quite a complex step, and it is explained in detail in this chapter.

A. Badia, *SQL for Data Science*, Data-Centric Systems and Applications,
https://doi.org/10.1007/978-3-030-57592-2_2

2.1 Database Tables

Relational databases store data in units called *tables*. In fact, a database can be considered a collection of tables. In almost all cases, such tables are linked to each other, in that they contain related or connected data. How this connection can be expressed is discussed later in Sect. 2.2.

SQL implements the concept of table in a straightforward manner. To create a table, the SQL language has a command, called (not surprisingly) `CREATE TABLE`. A very simple example of this command is

```
CREATE TABLE Employees (
  name char(64),
  age int,
  date-of-birth date,
  salary float)
```

In order to explain the meaning of this command, we need to give a description of a table's four components: the *name*, the *schema*, the *extension*, and the *primary key*.

Each table must have a name. Since a database may contain several tables, this name must be unique inside a database (tables in different databases can have the same name). As in the case of databases, this name is simply a label to be able to refer to the table, but it is customary to give the table a descriptive name that refers to whatever the data in the table denotes. Typically, tables have names like 'Employees,' 'country-demographics,' or 'New-York-flights.' This is the first component of the `CREATE TABLE` command.

The name of the table is followed by a description, in parentheses, of the table schema; this is a list of *attributes*. Each attribute refers to a column in the table (some books use 'attribute' and 'column' interchangeably). An attribute has a *name* and a *data type*. The name must be unique among all the attributes in the table. As in the case of tables, it is customary to give attributes meaningful names that evoke the kind of information or domain that they represent.

2.1.1 Data Types

A data type refers to the way in which a computer represents data values. All relational database systems have a repertoire of data types to choose from when defining a table. Even though each database system uses slightly different types with slightly different names, they are all very similar and easy to understand. Each system has the following basic types:

- *strings*: strings are the name given in Computer Science to sequences of characters. They are used to represent values from a discrete domain; usually, this is the data type for any nominal/categorical attribute and may also be used for ordinal types (see Sect. 1.3). Most systems will take any sequence made up of any combination of letters and number that does not start with a number (some systems may even admit labels that start with numbers). Thus, `rain_amount` or `DB101` are legal strings, but in some systems `4thterm` or `2011` may not be. Later on, when entering values (see Sect. 2.4) we will see that some systems require strings to be surrounded by quotes (single or double). The single quote is the standard SQL string delimiter.[1] Doing this allows the system to accept additional characters like whitespaces; this way the string 'Jim Jones' (or "Jim Jones") can be accepted as a single value.

Strings are also needed to give names to database elements, like tables and attributes on them; these names are called *identifiers*. Both MySQL and Postgres allow identifiers with and without quotes; however, when not using quotes, certain restrictions apply: an identifier cannot use hyphens or start with a digit or an underscore or use a reserved SQL word (like CREATE, ATTRIBUTE, etc.). When using quotes, these restrictions are all lifted.

The *size* of a string is the number of characters that make it up. Strings can be of several types, depending on their sizes and on how sizes are handled. With respect to the former, some strings are appropriate for short strings, like many categorical variables use; others are useful for large strings, like short snippets of text. With respect to size handling, some strings have *variable* size (called `character varying(n)` in SQL, with `n` a positive integer), others have *fixed* size (called `character(n)`). When a string value is shorter than the maximum size `n`, variable size strings adopt the size of the actual value, while fixed size strings are padded with whitespace (up to the maximum `n` characters). When the string value is longer than `n`, the value is truncated in either case. As an example, if we declare an attribute `CountryA` as `character varying(5)` and `CountryB` as `character(5)`, the value "USA" would be stored as such in `CountryA`, and as "USA " (note the two added whitespaces) in `CountryB`, while the value "Australia" would be truncated to "Austr" in both cases. When using a string data type, it is customary to examine existing data values for the attribute and choose a size that would allow even large values to be stored (with a few extra characters just in case). Many systems easily support strings of up to 256 characters, which is fine for most categorical variables (but may be short for certain uses. See Sect. 2.3.3 for handling long strings).

- *numbers*: used to represent numerical values, be they interval, ratio, or absolute (see Sect. 1.3). Most systems support several types of numbers: *integers* (called `int` or `integer` in most systems) for whole numbers and *float* or *decimal* for reals (rationals are expressed as reals). It is important to note, for those who are going to make heavy use of numerical calculations, that all numbers in a computer are expressed using a certain level of precision: for integers, this means that some numbers may be too large in absolute value; for real numbers, it means that some numbers may be too large or too small. In many systems, it is possible to specify a certain level of precision for real numbers, by using the notation

[1] In MySQL, double quotes work as a string delimiter by default.

`decimal(precision, scale)`: here, precision refers to the maximum (total) number of digits in the number, while scale refers to the maximum number of decimal places. For instance, the number 1234.567 has a precision of 7 and a scale of 3. Each system has a different limit on how precise numbers can be. In Postgres, the `double precision` data type offers 15 digit precision, and `float8` offers 17 significant decimal digits, while the type `bigint` can store integer values between -2^{63} and $+2^{63}$. MySQL offers similar types, with the same names and ranges. In both systems, the use of floating-point values may result in unexpected behavior when doing arithmetic with very large or very small values; we mention this in several relevant places in Chap. 3.

- temporal information types: usually, *date*, *time*, *timestamp*, and `interval` types are supported. A date is a combination of year, month, and day of the month that denotes a certain calendar day. A time is composed of hour-minute-second parts (some systems allow times with further precision, going down to tenths and hundreds of second). A timestamp is a combination of a date and a time, with all the components of both. Finally, an interval is a pair of timestamps intended to denote the beginning and end of a time period.

In addition to these, pretty much all systems have other types, like *Booleans* (to represent value True and False). In particular, many systems offer special types to store complex information. We will discuss types to store XML, JSON, and text data later in Sect. 2.3.[2]

It is easy to see that the same value could be represented by several data types. This leads to an interesting issue: we can pretty much declare anything as a string. For instance, one could declare an attribute `age` to be of string type in order to enter values like "almost 35" or "sixty-two" or "24."[3] However, as we will see in Sect. 3.1.2, the database provides several ways to manipulate each data type, represented by *functions*. Such functions are tailored to the expected values of the type. For instance, you cannot add two strings—but you can add two numbers. What this means is that, if this `age` attribute is declared as a string, we could not do simple arithmetic on it, even though it represents a numerical value. This would deprive us of the opportunity of even elementary analysis, like calculating the mean (average) of all the ages in our table. In order to do so, we should declare `age` as an integer; in general, we should always declare an attribute with the type that is most faithful to the values it contains. As noted in Sect. 1.3, one must take into account the semantics of the attribute: just because something looks like a number, it does not mean that it is a number (recall the example of the zip code; such value is better represented by a string).

---

[2]Even more advanced types exist that are able to store collections of objects. We will not be discussing them in this book, since they are usually not needed for data analytics and they add quite a bit of complexity to data modeling.

[3]Note that this is the string "24," not the number 24; these two are completely different.

Example: CREATE TABLE Statement

To create a table for the ny-flights dataset, we can use a command like the following:

```
CREATE TABLE NY-FLIGHTS(
flightid int,
year int,
month int,
day int,
dep_time int,
sched_dep_time int,
dep_delay int,
arr_time int,
sched_arr_time int,
arr_delay int,
carrier char(2),
flight char(4),
tailnum char(6),
origin char(3),
dest char(3),
air_time int,
distance int,
hour int,
minute int,
time_hour timestamp);
```

In choosing the data types, we have observed some sample values to determine an appropriate choice. Several aspects of this example are worth pointing out:

- Information about the date of the flight is expressed by 3 attributes, `year`, `month`, and `day`. Why not express it as a single attribute of type date? Many times, the *granularity level* at which to express the data is decided for us (as in this case). As we will see in Sect. 3.1.2, we can divide a string into parts fitting into a pattern, and we can also put together several strings into a single one. In these cases, it is easy to extract parts from the original value and to *concatenate* (put together) strings, so we can create an attribute of type date that contains the same information. If accessing the parts of a value is going to be done repeatedly, it may be a good idea to split the data in the first place. If breaking the value into parts is not an easy task (for instance, in real life dealing with dates is usually much more complicated), it may also be a good idea to break the attribute into parts, so that accessing the right data does not become a nightmare. Conversely, if the original data comes with the value as one big string, we may have to take it in as a single value and break up this value into parts with some database tools later. The right choice depends mainly on two circumstances: the raw data and how we are going to use the data.
- Attributes like departure or arrival time are given an integer data type. This is because that is how it is expressed in the raw data, with values like "517," instead of a time with hours and minutes. Again, if we are going to do arithmetic on those, it would be a good idea to have them as time values (note that as numbers, the difference between 517 and 455 is 65, but as times the difference is 22 min).

We cannot declare the attribute as time because the system cannot transform the raw values to time values, but we can convert between the integer value and a time value (the details of this transformation are explained in Sect. 3.3.1.3).
- attributes `carrier`, `flight`, `tailnum`, `origin`, and `dst` are given a fixed length. This is because most of these are codes (for instance, `origin` and `dst` are expressed as airport codes, which are 3 letters). If this were not the case, it would be better to go with a varying size attribute.
- the values of attribute `flight` look like numbers ('1545,' '1714') but are really codes, hence the choice of a string for data type. Again, ask yourself *how am I going to use this data?* (i.e. will I ever add two flight numbers?).
- the values of attribute `time_hour` are real timestamps (i.e. a date and a time) and therefore need to be declared as such so that the database can make sense of these values.

Exercise 2.1 Create a database and execute the CREATE TABLE statement for dataset `ny-flights` in both Postgres and MySQL.

2.1.2 *Inserting Data*

Finally, a part of the table that is not seen in the `CREATE TABLE` statement is the extension of the table, which contains the data in *rows* (AKA *tuples*). A row represents a record, object, or event; it gives a sequence of data values, one value for each attribute in the schema. In SQL, we first create the table; once, this is done, we can add rows to it, hence putting data into our database. This is done through the `INSERT` statement; the syntax is

```
INSERT INTO table-name VALUES(...)
```

The parentheses must enclose a list of data values that respects the schema of the table. Note that, since each value must match some attribute in the schema, we must pair values with attributes. This is done in one of the two ways:

- simply enumerating the values will pair them up with attributes following the order used in the `CREATE TABLE` statement (the default order). That is, if we created table T declaring (in this order) an attribute A of type integer, an attribute B of type string, and an attribute C of type date, the system will expect rows of the form (integer, string, date).
- specifying a customer order in the INSERT statement by following the table name with a list, in parentheses, of the attributes in the table, in whichever order we want. The values will then be expected to follow this order. For instance, reusing our previous example, we could use the statement

```
INSERT INTO T(B,C,A) VALUES(b,c,a)
```

where value b would be expected to be of type string (to correspond to attribute B), value c would be expected to be of type date (to correspond to attribute C), value a would be expected to be of type integer (to correspond to attribute A).

Example: Inserting Data

The statement

```
INSERT INTO NY-FLIGHTS VALUES(1, 2013, 1, 1, 517, 515, 2, 830,
819, 11, "UA", 1545, "N14228", "EWR", "IAH", 227, 1400, 5, 15,
2013-01-01 05:00:00);
```

is legal once the table NY-FLIGHTS has been created (we are using the 'default' order here). Observe that string values use double quotes.

Exercise 2.2 Insert two rows into the table ny-flights of the previous exercise. NOTE: some systems allow you to insert more than one row with a single INSERT statement; find out how to do that in Postgres and/or MySQL.

Note that the only data that can be *inserted* into the table is data that respects the schema, that is, rows that have exactly one value, and of the right type, for each attribute in the schema. It is sometimes the case that data does not *exactly* follow the schema, a problem that we examine in more detail later. SQL offers some flexibility for the cases where we have *incomplete* data (some values are missing): a special marker, called the NULL marker, can be used to signal that a value is missing.

Example: Inserting Incomplete Data

Assume that a row of NY-FLIGHTS is missing the departure time, the scheduled departure time, the tail number, and the distance. Then we can use the following statement to insert the row:

```
INSERT INTO NY-FLIGHTS VALUES(2, 2013, 1, 1, null, null, 4,
850, 830, 20, "UA", 1714, null, "LGA", "IAH", 227, null, 5,
29, 2013-01-01 05:00:00);
```

It is also possible to destroy a table by using the SQL command
DROP TABLE tablename
This will get rid of the table *and* of any data in the table.
A few final observations about tables:

- for simplicity, most tables in this book have very simple schemas, and only a small portion of their data is shown. Real-life tables may have from a few to a few hundred attributes in the schema and may have hundreds to thousands or millions of rows in the extension. Since it may be necessary to remember what attributes exist in a table, and what their exact name is, all systems provide a method to retrieve information about a table, including its schema. In MySQL, there is a command

```
SHOW COLUMNS [FROM table_name] [FROM database_name]
```

that will give schema information about the (optional) `table_name` argument. The same information can be retrieved in Postgres at the command line using

```
\dt *.*
```

for all tables in all databases, or

```
\dt database_name.table_name
```

for a particular table in a particular database. It is also possible to retrieve this information directly, since all relational databases store metadata (data about the databases) in tables! In MySQL, this is done with the statement

```
SELECT COLUMN_NAME, DATA_TYPE
FROM INFORMATION_SCHEMA.COLUMNS
WHERE table_name = 'table_name'
    [AND table_schema = 'database_name'];
```

A similar query will also work in Postgres. Queries (SELECT statements) are explained in detail in the next chapter.

- As a rule, the schema of a table changes very rarely; the rows of data may change rapidly. As we will see, it is very easy to change the schema of a table before it contains data, but it becomes problematic once data is on it. This is one reason that schema changes are infrequent. However, they are possible, as discussed in Sect. 3.5.
- Even though database tables can be depicted as described in the previous chapter, it is important to note that the order of the columns or the rows is completely irrelevant. In other words, a table with the same number, type, and names of columns in a different order would be considered to have the same schema. And two tables over the same schema with the same rows, even in different order, would be considered to have the same data.

2.1.3 Keys

Among all attributes that make up the schema of a table, it is common that some of them together are enough to identify rows, that is, to tell a row apart from all the others.[4] Such attribute sets are called *keys* in databases (in other contexts, they

---

[4]Technically, this is *guaranteed* to be the case, since all attributes together (the whole schema) could be a last-instance key. But in most cases, a small subset of attributes works.

are called *identifiers*, and the rest of the attributes, *measurements*). When creating tables, we identify and declare keys for each table.

Example: Keys in Tables

In the table `ny-flights`, a key is a combination of attributes that identifies a flight (i.e. distinguish a certain flight from all others). There is an obvious key (the first attribute, `flightid`). There are also several other keys: a combination of departure date (`year, month`, and `day`), `carrier` and `flight` number will also identify each flight uniquely (unless an airline has the same flight that departs several times a day under the same flight number!). Note that the tail number (which identifies the airplane) may also be used instead of flight number if each flight is assigned a certain plane, so that departure date, `carrier`, and `tailnum` may also be a key (there is no rule forbidding two keys from sharing attributes!). There may be other keys, like a combination of departure date, `dep_time`, `origin`, and `dest`. Whether any of these is a true key depends on the exact semantics of the attributes.

It is important to understand two things about keys:

- In SQL, a table can have *repeated rows* (that is, the same row can be added to a table several times). Obviously, repeated rows have the same key. Thus, when we say that a key can tell a row from any other, we mean "from any *distinct* row." A key value is always associated with one or more copies of a given row. What should never happen is that a key is present in two (or more) *different* rows. In that case, what we have is not really a key.
- a key is a group of attributes that is *as small as possible*—that is, all the attributes in the key are needed for the key to do its job of telling rows apart. Without this stipulation, we would have a large number of (useless) keys. Consider, for instance, a table with information about people, where one attribute is the *social security number* of the person. Since this is (at least in the USA[5]) a unique code per person, we can think of this attribute as a single-attribute key. But then, technically, the pair of *social security number* and *name* (or any other attribute) could also work as a key, since given a *social security number* and a *name*, we can tell any two rows apart. The problem, of course, is that *social security number* is doing all the work; *name* is just along for the ride. In fact, without the requirement of minimality, any set of attributes that included *social security number* would be a key. So when we talk about a key we will always mean a *minimal* set of attributes.

Finding keys is very easy sometimes and far from trivial in others. We may have situations in which a table has more than one key. In such a scenario, we choose one key to be the *primary key* of the table; all other keys are called *unique* attributes

[5]Other countries also give their citizens some sort of identification number.

in SQL. Sometimes it is unclear what constitutes a key for a table; it has become customary in many scenarios to *create* artificial keys, called *identifiers* (*id* for short). Usually, an id is simply an integer that is assigned to each row by giving a 1 to the first row inserted in the table, a 2 to the second row, etc. This guarantees that each row has a different value; hence, an id works perfectly as a key (we will see an example of how this is done in SQL later on). There are two things worth mentioning about ids: first, they are meaningless. They technically satisfy all the requirements of a key, but (unlike other attributes) do not refer to anything in the real world. Therefore, they are useless for most data analyses. Second, if there is a key in a table, and we decide to create an id, the key is still there, and if it goes undetected (so it is not even declared as UNIQUE) this may lead to some problems later on. The moral of the story is that we should not rush to create ids; rather, we should examine the existing attributes carefully to identify any possible keys.

Example: Ids and Multiple Primary Keys

In the case of `ny-flights`, the attribute `flightid` is an example of id, and we have multiple (possible) keys. As another example, suppose we have a dataset about cars, where some of the attributes are: *VIN*, or Vehicle Identification Number, a unique number given to each car on the road in the USA; *State*, the state where a vehicle is registered; *license-number*, the license (plate) number for a card; *make*, or manufacturer; *model*; and *year*.

Vin	State	License-number	Make	Model	Year
WBAA3185446N384	KY	444ABC	BMW	325i	1989

We can tell that:

- *VIN* is, by itself, a key. This is due to the fact that it is designed to work as such; a different VIN is assigned to each car.
- *State* and *License-number* are, together, another key. This is because each state gives each car licensed in that state a unique plate. Thus, two cars from different states could end up with the same place, but no two cars from the same state can have the same license number.

In this case, we have two keys. We choose one as primary and declare the other one as unique. It is customary (but not required) to choose as primary the simplest (smallest) key, so we would choose *VIN* as primary and *State, license-number* as unique.

Note that whether a group of attributes is a key depends entirely on the semantics of the attributes in the schema. If a group of attributes is a key, it should work as a key for any possible extension of the table, not just from the data that we have right now. Hence, it is not possible to determine whether a group of attributes is a key by inspecting the data. It is, however, perfectly possible to prove that a group of attributes is *not* a key by finding a counter-example in the data.

As we will see shortly, primary keys fulfill an important mission: they allow us to refer to specific tuples in a table. This will become necessary when organizing complex datasets that require more than one table (see Sect. 2.2) and when making changes to the data (see Sect. 2.4.2).

Example: Declaring Keys

We can declare keys (and uniques) when creating a table; for instance, in Postgres, we could have used

```
CREATE TABLE ny-flights(
  flightid int PRIMARY KEY,
  ...);
```

in our original CREATE TABLE statement—or, alternatively, we could add a line after the last attribute:

```
CREATE TABLE ny-flights(
  flightid int,
  ...
  time_hour timestamp,
  PRIMARY KEY (flightid));
```

instead of our original example. We can also modify an existing table to add information about keys. For instance, in Postgres we could create table `ny-flights` as before and then issue the SQL command:

```
ALTER TABLE ny-flights ADD PRIMARY KEY(flightid);
```

Technically, the declaration of a primary key is what is called a *constraint* and it can also be written as such, by using the keyword `CONSTRAINT` and giving a name:

```
CREATE TABLE ny-flights(
  flightid int CONSTRAINT nyf-pk PRIMARY KEY,
  ...);
```

or as

```
CREATE TABLE ny-flights(
  flightid int,
  ...
  time_hour timestamp,
  CONSTRAINT nyf-pk PRIMARY KEY (flightid));
```

This gives a name (`nyf-pk`) to the primary key declaration. The syntax is the same in MySQL and in all relational systems, as it is part of the SQL standard.

Exercise 2.3 Alter the `ny-flights` table in Postgres and MySQL to add a primary key to it.

Example: Creating Ids

As stated earlier, most systems have some way of creating id attributes. In Postgres, an attribute is given type SERIAL and declared a primary key, and the system automatically creates values for this attribute:

```
CREATE TABLE books (
  book-id   SERIAL PRIMARY KEY,
  title    VARCHAR(100) NOT NULL,
  author  VARCHAR(100),
  publisher VARCHAR(100),
  num-pages INT);
```

In this case, when adding data into the table it is not necessary to provide a book-id; the system will automatically generate a new value for each row. That is, insertions into books will provide rows with only 4 values: one for title, one for author, one for publisher, and one for num-pages.

In MySQL, the same example is written as follows:

```
CREATE TABLE books (
  book-id    INT NOT NULL AUTO_INCREMENT PRIMARY KEY,
  title    VARCHAR(100) NOT NULL,
  author  VARCHAR(100),
  publisher VARCHAR(100),
  num-pages INT);
```

Note that an id primary key can also be added to a table after creation; if we had not declared book-id as above, we could add

```
ALTER TABLE books
ADD book-id INT AUTO_INCREMENT PRIMARY KEY;
```

in MySQL, and similarly in Postgres:

```
ALTER TABLE books
ADD book-id SERIAL PRIMARY KEY;
```

Other keys beside the primary one can be declared exactly like the primary key, but using the keyword UNIQUE instead of PRIMARY KEY.

Example: Unique Keys

For the table CAR described previously, we can use the declaration

```
CREATE TABLE CAR (
 Vin VARCHAR(36) PRIMARY KEY,
 State CHAR(2),
 License-number CHAR(6),
 Year INTEGER
 UNIQUE (State, License-number)
)
```

The last line could also be

```
CONSTRAINT uniq-car UNIQUE (State, License-number)
```

Likewise, ALTER TABLE can also be used to add a unique constraint after the fact:

```
ALTER TABLE car ADD UNIQUE (State, License-number)
```

A primary key cannot have NULL markers in any of its attributes; if we declare a set of attributes as the primary key of a table, any attempt to enter a row with null markers on any part of the key will be automatically rejected by the system. This is because the system will check that a primary key (or another key declared as UNIQUE) is indeed a (primary) key; if a (primary) key contains nulls, the system cannot ascertain whether it is a correct key or not. If we declare a (primary) key and misuse it (by entering the same value of the so-called key in two distinct tuples), the system will reject the insertion because it violates the condition of being a key.

2.1.4 Organizing Data into Tables

A large proportion of data used for data analytics is presented in a tabular format. In fact, many algorithms for Data Mining and Machine Learning assume that data is in a table. This is why it is important to understand what a proper table is, and how to deal with data that is not in the right format.

We start with the basic observation that the same data can be structured in several ways. Not all of them are conducive to analysis.

Example: Wide (Stacked) and Narrow (Unstacked) Data

We have some data about people, presented in two different ways.[6]

Person	Age	Weight
Bob	32	128
Alice	24	86
Steve	64	95

Person	Variable	Value
Bob	Age	32
Bob	Weight	128
Alice	Age	24
Alice	Weight	86
Steve	Age	64
Steve	Weight	95

[6] This example is a simplification of the one used in Wikipedia at `en.wikipedia.org/wiki/Wide_and_narrow_data`.

The first table is called a *wide* or *unstacked* table, while the second table is sometimes called a *narrow* or *stacked* table. For the purpose of analysis, narrow/stacked tables are a very inconvenient structure; we want our tables always in the wide format. Note that in the wide table the key is `Person`, suggesting that each row describes a person, while in the narrow table the key is (`Person, Variable`), a strange combination that is a hint that something is not quite right.

We will see later that there are ways to transform data from narrow (stacked) to wide (unstacked) and vice versa. Data in narrow format is, unfortunately, a common occurrence.

There are other ways in which data may be presented in an undesirable format.

Example: Untidy Data

A very common situation is to have data as follows:

Product	S	M	L
Shirt	12.4	23.1	33.3
Pants	3.3	5.3	11.0

This is undesirable, since the sizes S (small), M (medium), and L (large) can be considered the values of a categorical attribute but are used in the schema. This makes some analysis rather difficult. The following arrangement works better:

Long table

Product	Size	Price
Shirt	S	12.4
Shirt	M	23.1
Shirt	L	33.3
Pants	S	3.3
Pants	M	5.3
Pants	L	11.0

Again, the first table is an example of narrow/stacked data, while the second is an example of wide/unstacked data. A proper table data (called *tidy data* in [19]) is wide/unstacked; to achieve this, it follows certain conventions:

- all values are in the data cells, never in the schema. For instance, S (small), M (medium), and L (large) are values of attribute `size`, so they should be in data cells, not part of the schema.
- all attributes are in the schema, never in the data. For instance, `Age` and `Weight` are attributes of a person, so they should be in the schema, not in the data.
- each row corresponds to a data record (observation, in statistics) and each column to an attribute (variable, in statistics). Intuitively, in a table about people we want

each row to describe a person; in a table about experiment runs, each row should be a run. Note that, in the last example above, each row describes a combination of product and size; this is due to the fact that each product is sold in several sizes. We will see how to deal with this situation shortly.

Having untidy data is a common problem; both R and Python provide tools to deal with this situation.[7] However, both R and Python are a bit more liberal than databases about the data they admit. The following provides an example of what pandas accept that a database does not.

Example: Tidy Data in Other Frameworks

The following tables provide an example of how the panda framework would unstack/pivot a data frame (the name of the structure in pandas that holds tabular data). The dataset on the left is a stacked data frame, the one on the right is an unstacked data frame.

		Spring	Fall
Emma	History	82	91
	Physics	81	79
Gabi	History	80	88
	Physics	83	89

	Spring		Fall	
	History	Physics	History	Physics
Emma	82	81	91	79
Gabi	80	83	88	89

However, we would not call the data frame on the left a proper table in SQL (and it would not be tidy according to [19]). The proper table would look as follows:

Student	Subject	Semester	Score
Emma	History	Spring	82
Emma	Physics	Spring	81
Emma	History	Fall	91
Emma	Physics	Fall	79
Gabi	History	Spring	80
Gabi	Physics	Spring	83
Gabi	History	Fall	88
Gabi	Physics	Fall	89

Recall that in a table the order of rows (or attributes) is irrelevant. On the data frame, the 'location' of a score is crucial; in the table, it is the values in the tuple that matter. The tuple itself can be anywhere in the table.

The problem of ill-structured data is very common in spreadsheets. It is compounded there by the fact that data and analysis are often mixed up. The following is an example of the problems found when data comes from spreadsheets.

[7] R has a `tidyr` package; Python provides several functions on the Panda library.

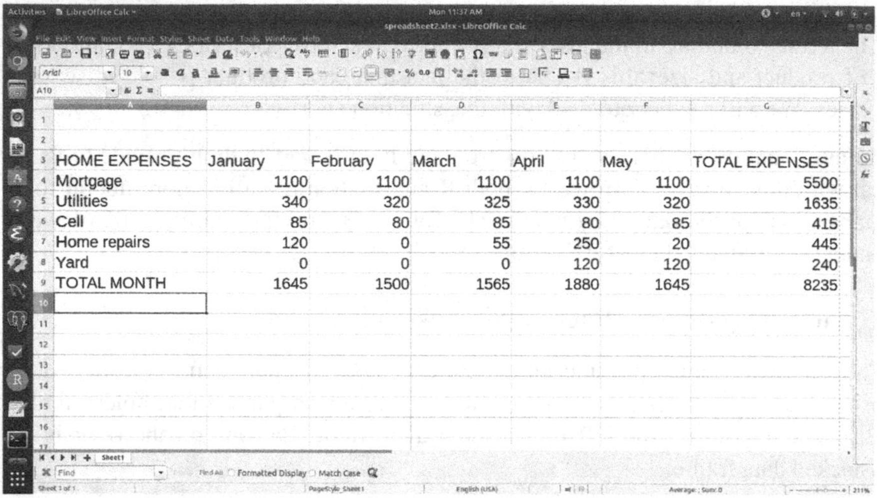

HOME EXPENSES	January	February	March	April	May	TOTAL EXPENSES
Mortgage	1100	1100	1100	1100	1100	5500
Utilities	340	320	325	330	320	1635
Cell	85	80	85	80	85	415
Home repairs	120	0	55	250	20	445
Yard	0	0	0	120	120	240
TOTAL MONTH	1645	1500	1565	1880	1645	8235

Fig. 2.1 Non-tabular data in a spreadsheet

Example: Spreadsheets Problems

Another typical arrangement of data that is not tabular can be seen in Fig. 2.1. As before, the problem is that the data has been put on both down the columns and across the rows (the date when an expense was incurred is on the top row). In a tabular (tidy) format, this data would have schema *(Expense, Month, amount)*, with data like

Expense	Month	Amount
Mortgage	January	1100
Utilities	January	340
	...	
Mortgage	February	1100
	...	

There is an additional issue here. The bottom row and the rightmost column are what is usually called *margin sums*; they are data derived by calculation from the original data. When bringing the information in the spreadsheet into a table, there is no sense in copying these derived data; these results can be easily calculated in SQL, and they are different in nature from the raw data in the table.

Sometimes the problems with spreadsheets go deeper than this. The root of the problem is that spreadsheets do not enforce any rules on how data is organized. As a result, a user can do whatever seems convenient for a given situation. For instance, if we look at the spreadsheet in Fig. 2.2, which is a generic template from Google docs, we will see that part of it looks like the table in the previous example, but in addition there is also other data (the attributes at the top: 'Employee Name,' 'Company Name,'...) that is different from the data in the table. Another common situation arises when a spreadsheet has data spread across multiple pages or tabs. Suppose, for instance, that we have a dataset about properties in a spreadsheet, which each property described in a separate page. There is no guarantee that the same or even similar data is present on each sheet or that it is in the same format. And even though the data is in separate pages, it is all part of the same datasets. In many cases, all the data in the spreadsheet can be put into a database, sometimes

WEEKLY TIMESHEET TEMPLATE

EMPLOYEE NAME: ____________ COMPANY NAME: ____________

EMPLOYEE ID: ____________ COMPANY ADDRESS: ____________

MANAGER NAME: ____________

WEEK BEGINNING: Sunday, April 30, 2017 COMPANY PHONE: ____________

DATE	REGULAR HRS	OVERTIME HRS	SICK	VACATION	HOLIDAY	OTHER	TOTAL HOURS
04-30-2017	8.00	2.25					10.25
05-01-2017						8.00	8.00
05-02-2017	8.00						8.00
05-03-2017	8.00						8.00
05-04-2017	8.00	4.50					12.50
05-05-2017					8.00		8.00
05-06-2017	4.25						4.25
TOTAL HOURS	36.25	6.75	0.00	0.00	8.00	8.00	59.00
RATE PER HOUR	$20.00	$30.00	$20	$20	$20.00	$20.00	
TOTAL PAY	$725.00	$202.50	$0.00	$0.00	$160.00	$160.00	$1,247.50

BI-WEEKLY TIMESHEET TEMPLATE

EMPLOYEE NAME: ____________ COMPANY NAME: ____________

EMPLOYEE ID: ____________ COMPANY ADDRESS: ____________

MANAGER NAME: ____________

WEEK BEGINNING: Sunday, April 30, 2017 COMPANY PHONE: ____________

DATE	REGULAR HRS	OVERTIME HRS	SICK	VACATION	HOLIDAY	OTHER	TOTAL HOURS
04-30-2017	8.00	2.25					10.25
05-01-2017			8.00				8.00
05-02-2017	8.00						8.00
05-03-2017	8.00						8.00
05-04-2017	8.00	4.50					12.50
05-05-2017					8.00		8.00
05-06-2017							0.00
05-07-2017							0.00
05-08-2017	6.00					8.00	14.00
05-09-2017	8.00	3.75					11.75
05-10-2017	8.00						8.00
05-11-2017	8.00	4.50					12.50
05-12-2017					8.00		8.00
05-13-2017							0.00
TOTAL HOURS	30.00	8.25	0.00	0.00	8.00	8.00	54.25
RATE PER HOUR	$20.00	$30.00	$20	$20	$20.00	$20.00	
TOTAL PAY	$600.00	$247.50	$0.00	$0.00	$160.00	$160.00	$1,167.50

Fig. 2.2 Spreadsheet example

in a single table, sometimes in multiple tables (we discuss this example again in Sect. 2.2).

We can, with some discipline, use spreadsheets to represent tabular data, but we can also do all kinds of weird stuff to our data—one of the several reasons for not using spreadsheets for serious data analysis.[8]

Sometimes data presents multiple problems at once. In fact, massaging the data into the right format for analysis is usually a (time-consuming and) necessary step before any analysis is done (we discuss this in Sect. 3.4.1).

Example: Very Untidy Data

The United Nations has several datasets publicly available at `data.un.org`. One such dataset shows international migrants and refugees; when downloaded as a csv file, it looks like this:

Region/Country/Area		Year	Series	Value
1	Total	2005	International migrant stock: Both sexes (number)	190531600
1	Total	2005	International migrant stock: Both sexes (% total population)	2.9124
1	Total	2005	International migrant stock: Male (% total population)	2.9517
1	Total	2005	International migrant stock: Female (% total population)	2.8733
1	Total	2010	International migrant stock: Both sexes (number)	220019266
1	Total	2010	International migrant stock: Both sexes (% total population)	3.162
1	Total	2010	International migrant stock: Male (% total population)	3.2381
1	Total	2010	International migrant stock: Female (% total population)	3.0856

This dataset has several problems: it clearly is narrow/stacked; if we go down looking at the second (unnamed) attribute, we find values `Total` (shown above), `Africa`, `Northern Africa`, `Algeria`,... Clearly, the data is shown in several levels of granularity. This really corresponds to what we call later *hierarchical data* (see Sect. 2.3.1) and should not be mixed up in the same table. Besides pivoting, this data needs also to be separated into different tables.

2.2 Database Schemas

Up until now, we have assumed that data fits in a 'tabular' format. Basically, this means that

- each record (event/experiment/object) in our domain has a well-defined set of attributes (features/variables/measurements), which are the same for each record (that is, the record set is homogeneous). For instance, in the table `Books` each

[8]There are other good reasons, but we will not discuss them in this book. After all, we have only so many pages.

row represents a book; in the table `ny-flights` each row represents a flight, and so on.
- each such attribute has exactly one value per record. For instance, in a table about People, each person has exactly one age, hence one value for attribute `Age`.

But sometimes data does not follow these rules. We now examine the most common reason for data not fitting well in a tabular format, and what can be done about it (in this section, we still assume structured data. Semistructured and unstructured data is dealt with in Sect. 2.3).

There are three common scenarios for data not to be 'tabular':

- heterogeneous data, that is, data where records may have values for different attributes;
- data with multi-valued attributes, that is, attributes that have more than one value for a record; and
- complex data involving different (but related) events or entities.

We study each one next.

2.2.1 Heterogeneous Data

In the first scenario, we may have records in our dataset that are not completely homogeneous: while sharing many common attributes, some entities may be somewhat different. This may be due to some attributes being present only under certain circumstances and not applying to each record (these are sometimes called *optional* attributes). For instance, an attribute `spouse-name` only makes sense when the record is about a married person, but would not have a value for single people. In this case, we have two options:

1. create a single table and add all attributes present in any record to the schema. Then, on each row, we describe one record; when the record does not have a value for an attribute, we use a NULL marker. This option keeps the database simple, but at analysis time all the NULLs need to be accounted for (we discuss this in Sect. 3.3).
2. identify sets of records that share the same attributes and create a table for each. This leads to *fragmentation* (the same dataset is represented by several tables, not just one) but eliminates the problem with NULLs. Clearly, this option is only advisable if all data can be divided into a few groups of homogeneous attributes; fortunately, this is not an uncommon situation in real-life datasets.

Example: Optional Attributes

In the previous chapter, we used the Chicago employee dataset as an example of tabular data. The schema of the table had attributes *Name, Job Title, Department,*

Full or Part Time, Salary or Hourly, Typical Hours, Annual Salary, and Hourly Rate. We noted that there are missing values in each record. This is pretty much guaranteed by the fact that the table includes both salaried and hourly employee (as recorded in attribute *Salary or Hourly*): salaried employees will have an *Annual Salary* but no *Hourly Rate*, while hourly employees will have an *Hourly Rate* but not *Annual Salary*. The problem with this table is that it is going to have a large number of NULL markers. However, if we are going to run an analysis that involves all the employees, we may want to keep them in a single table:

Employee							
Name	**Job Title**	**Department**	**F/P**	**S/H**	**TH**	**AS**	**HR**
Jones	Pool Motor Truck	Aviation	P	Hourly	10	Null	$32.81
Smith	Aldermanic Aide	City Council	F	Salary	Null	$12,840	Null

(again, we have abbreviated the attribute names). If we want to avoid any missing data, then we should split this table into two, one for salaried employees and one for hourly employees, and use only the relevant attributes on each (note that we can make do without attributes `Salary/Hourly` since each table is only for one type of employee).

Hourly Employee					
Name	**Job Title**	**Department**	**Full/Part Time**	**Typical Hours**	**Hourly Rate**
Jones	Pool Motor Truck	Aviation	P	10	$32.81

Salaried Employee				
Name	**Job Title**	**Department**	**Full/Part Time**	**Annual Salary**
Smith	Aldermanic Aide	City Council	F	$12,840

We can go back and forth between the two designs; SQL allows combining the two tables together into a single table design (this is explained in Sect. 5.4) as well as breaking down the single, original table into two separate ones (as shown in Sect. 3.1).

2.2.2 Multi-valued Attributes

In the second scenario, we have that some records are characterized by attributes that have several values for a given record (these are called *multi-valued attributes* in database textbooks). An example of this was seen in a previous example, where a record came in several sizes, each one with a price. This forces us to describe the record using several rows, since we can only put one value for the attribute on each row. This is fine and will not present a problem for analysis.

Example: Transactional Data

In business applications, it is common that we think in terms of 'transactions,' each one of each involves a set of products (usually, buying or selling transactions, with products being the products/services being bought or sold). To describe this data, we usually assign an identifier to each transaction and associate all involved products with the transaction they belong to. The following example describes a couple of sales in a store:

Products	
Transaction-id	**Product**
1	Bread
1	Beer
1	Diapers
2	Eggs
2	Milk

In this table, there are two transactions; the first one involves three products; the second one, two products. Because of this, the first transaction is shown in 3 rows; the second one spans 2 rows. Note that, even though `Transaction-id` is meant to identify the transaction, the key of this table is (`Transaction-id, Product`), since the key needs to be unique for each row.

This scenario is quite common; for instance, any situation where we take repeated measures over time leads to this situation (we discuss the influence of time in databases at the end of this section).

In many cases we have mixed situations, in which a record has both regular (single-valued) and repeated (multi-valued) attributes. Staying within the single table format in a mixed situation creates some problems.

Example: Transactional Data, Revisited

Assume that, for each transaction, we have some information that is unique to the transactions, like the date and time and the store. In that case, the strategy above would involve repeating the information concerning the single-valued attributes.

Transactions-and-Products				
Transaction-id	**Product**	**Date**	**Time**	**Store**
1	Bread	10/10/2019	10:55pm	Market-street
1	Beer	10/10/2019	10:55pm	Market-street
1	Diapers	10/10/2019	10:55pm	Market-street
2	Eggs	10/11/2019	8:25pm	Main-street
2	Milk	10/11/2019	8:25pm	Main-street

Repeating data in this manner may be fine for analysis purposes (it is common to have categorical or ordinal attributes repeated like this, as we will see), but it can present some serious trouble for other purposes. In particular, if we are storing the data and it is possible that we modify it later (by deleting or modifying existing data or adding more), we are facing what is called *anomalies*, situations where the database may be left in an inconsistent state by changes (inconsistency here means that data in some tuples contradicts the data in other tuples). A typical example is the *update anomaly*: suppose that the date of transaction 1 is wrong, and I want to change it. As we will see in Sect. 2.4.2, there is an `UPDATE TABLE` command that allows to change existing rows in a table. However, here I have to make sure that I update the 3 rows spanned by the description of transaction 1; if I change one or two of the rows only, I would leave a transaction associated with two dates. Under the assumption that each transaction happens in one and only one date, this means that the data in the database is inconsistent. But once the change has been made, it is impossible to decide which date is the correct one and which date is the incorrect one (unless we have access to the original record of sale). Hence, it is better to avoid the anomalies in the first place. The method to do so is called *normalization*, and it consists of spreading the data into several tables by separating the multi-valued attributes from the single-valued ones.

Example: Database Design

We are going to store the same information as in the previous example using normalization. First, we separate the table into two tables, one with the transaction information (single-valued attributes) and another one with the transaction products (multi-valued attribute).

Transactions			
Transaction-id	**Date**	**Time**	**Store**
1	10/10/2019	10:55pm	Market-street
2	10/11/2019	8:25pm	Main-street

Products	
Transaction-id	**Product**
1	Bread
1	Beer
1	Diapers
2	Eggs
2	Milk

Note that the first table has only two rows: a single row per transaction is enough. The second table is now back to our original example: the id of the transaction is repeated, but nothing else. In fact, if we count the number of cells in these two tables, we see that they have $8 + 10 = 18$ cells total. The original table in the previous example has 25 cells. And yet, these two tables have exactly the same information as the original one. The extra cells were created because of redundancy.

The key of the first table is `Transaction-id`, as it stores each transaction information in one row. This is possible because only single-valued attributes are used in the schema of the table. Conversely, the second table stores the multi-valued attribute `Product` and (as stated above) has a different key. However, it keeps a copy of `Transaction-id` so it can connect records to the transactions as described in the previous table. Copies of primary keys are used in databases to keep connections in data and are called *foreign keys*. This technique is crucial in database design, as we will see shortly.

As in the previous case, we can use SQL to transform a single table into several and to combine several tables back into a single one (this is discussed in detail in Sect. 3.1.1). The decision whether to normalize tables or not is, therefore, entirely a pragmatic one. The decision should be guided partly by what we intend to do with the data (are we at risk of having anomalies? That is, is the data going to be modified or just analyzed?) and partly by the structure of the data itself: as we will see shortly in another example, sometimes data has a complex structure and it is not a good idea to put it all in one table.

2.2.3 Complex Data

The strategy of normalization is also used in this third scenario: situation in which we must deal with complex data. This is the case where we have data about several related records (events and/or entities). For instance, we may have data about the customers of a business and the orders they have placed; we can think of customers and orders as different records, but they are related—since orders are placed by customers. As a different example, we may have a group of patients who participate in a drug study; during this study, each patient has several vital signs (blood pressure, etc.) measured daily for a period of time. The patient is a record; the daily measurements, an event—but clearly, the measurements and the patients are related. We may even have information about the study (who is in charge of it, when it started, etc.); the study is related to the patients and the measurements.

In general, we may have an arbitrary number of events/entities and connections between/among them. However, in most scenarios the connections among event/entities are *binary* (they involve two events/entities) and, for the purposes of database design, they can be classified in one of the three types:

- A collection of record I_1 has a *one-to-one* relationship with a collection of records I_2 if a record in I_1 is related to only one record in I_2 and vice versa (a record in I_2 is related to only one record in I_1). For instance, assume we have a demographic database with data about people, but also data about addresses (the exact longitude/latitude, state, etc.) and that each person is associated with one address and each address with one person.
- A collection of record I_1 has a *one-to-many* relationship with a collection of records I_2 if a record in I_1 is related to only one record in I_2 but a record in I_2

may be related to several records in I_1. For instance, in our example above of customers and orders, we can assume that some customers have placed several orders, but each order is associated with one and only one customer. Similarly, in the case of the clinical trial, each participant has several measurements taken, but each measurement is related to only one participant.

- A collection of record I_1 has a *many-to-many* relationship with a collection of records I_2 if a record in I_1 may be related to several records in I_2 and vice versa (a record in I_2 may be related to several records in I_1). For instance, assume we have a dataset of different chemical compound suppliers and another one of chemical laboratories. Each laboratory buys from several suppliers; each supplier sells to several laboratories: there is a many-to-many relationship between supply companies and laboratories.

In general, we want each event/entity to have its own table; this will allow us to tailor the schema of the table to those attributes that describe the event/entity in the most useful or meaningful way. For instance, a table for `Patients` may have attributes like `Name`, `date-of-birth`, `insurance-company`, etc., while a table for `Studies` may have attributes like `date-started`, `sponsor`, and so on. If an event/entity has single-valued and multi-valued attributes, then normalization (as shown in our previous examples) is traditionally used—although, as we have seen, whether to normalize a table or not is a design decision that must take into account the uses of the data.

Besides that, we have to capture the relationships between/among events/entities. The way such relationships are expressed in databases is by using the primary key of an event/entity as its surrogate: primary keys are copied to another table to represent the event/entity and express the relationships. In the previous example we saw that `TransactionID` is the primary key of the table describing transactions, but is also part of the schema (and part of the key) of the table that relates transactions with their records. Recall that a primary key copied to another table is called a *foreign key* in SQL (hence, in the second table above `TransactionID` is a foreign key). All relationships are expressed through foreign keys in relational databases; therefore, a foreign key must be declared to the database, just like a primary key.

Example: One-to-One Relationships

Assume, as before, that we have information about people (heads of households, really) and addresses, with each person having at most one address and each address belonging to one and only one person. Then we can express this simply combining all information in a single table:

Name	Age	...	Street-number	Street-name	...
"Jim Jones"	39	...	1500	Main-Street	...
"Fred Smith"	45	...	1248	Market-Street	...

There are other options for representing this information: we could have two separate tables (`People` and `Addresses`) and copy the primary key of one of them in the other (i.e. as a foreign key). It does not matter which key is copied; either one will do. This is done in SQL as follows:

```
CREATE TABLE PEOPLE (
Name VARCHAR(64) PRIMARY KEY,
,...)

CREATE TABLE ADDRESS (
street-number INT,
street-name VARCHAR(128),
city VARCHAR(64),
...
Name VARCHAR(64) FOREIGN KEY REFERENCES PEOPLE
PRIMARY KEY (street-number, street-name, city))
```

We have taken the (pragmatic) decision of using the primary key of table `People` as a foreign key in table `Address` because it is simpler than the primary key of `Address`. However, even a primary key with several attributes can be used as a foreign key; the whole key needs to be copied. For instance, we could have used the primary key of `Address` as a foreign key in `People` as follows:

```
CREATA TABLE PEOPLE (
Name VARCHAR(64),
,...
street-number INT,
street-name VARCHAR(128),
city VARCHAR(64),
...
PRIMARY KEY (Name),
FOREIGN KEY (street-number, street-name, city)
      REFERENCES ADDRESS)
```

For data analysis purposes, the single table option is usually the best in this case. For the next cases, the choice is not so clear-cut.

Example: One-to-Many Relationships

In the example used above, we have customers and orders, and each order corresponds to one customer, but a customer may place several orders. First, we would create a table `Customer`, with primary key `CustomerID` and table `Order`, with primary key `OrderID`, to hold information about customers and orders. Then we would add an attribute `CustomerID` to the schema of `Order`; on each row (for each order) we would add the id of the customer to identify who placed the order. That is, tables would be created as follows:

```
CREATE TABLE CUSTOMER(
CustomerID INT PRIMARY KEY,
...);

CREATE TABLE ORDER(
OrderID   INT PRIMARY KEY,
CustID INT FOREIGN KEY REFERENCES Customer,
...);
```

In this table, we may have data like the following:

Customer			
CustomerID	**Name**	**Address**	...
1	Wile E. Coyote	999 Desert Rd.	...
2	Elmer Fudd	123 Main St.	...

Order				
OrderID	**CustID**	**Date**	**TotalAmount**	...
300	1	9/12/2010	20,000	...
301	1	1/2/2011	15,000	...
302	2	2/5/2011	500	...
303	2	6/5/2012	800	...

If put together in a single table, the customer information would have to be repeated (we do not repeat the foreign key, as it is redundant in this table):

Customer-Order						
CustomerID	**Name**	**Address**	**OrderID**	**Date**	**TotalAmount**	...
1	Wile E. Coyote	999 Desert Rd.	300	9/12/2010	20,000	...
1	Wile E. Coyote	999 Desert Rd.	301	1/2/2011	15,000	...
2	Elmer Fudd	123 Main St.	302	2/5/2011	500	...
2	Elmer Fudd	123 Main St.	303	6/5/2012	800	...

Note how each customer information is repeated as many times as orders the client has.

A foreign key is declared, just like a primary key, by telling the database which attribute(s) form the foreign key and which table (hence which primary key) this foreign key is a copy of. Note that the type and number of attributes in the foreign key must match the primary key it copies: if the primary key is a single attribute of type integer (as in this example), the foreign key must be a single attribute of type integer. If the primary key were made of two attributes, one of type string and one of type date, any foreign key should also have two attributes, one of type string and one of type date. The reason for this is that any values in the foreign key must be values already appearing in the referenced primary key. For instance, in the example above, any `CustID` entered as part of a tuple of table `Order` must already appear in the primary key of table `Customer`. The database will enforce this rule by rejecting any data that does not obey it. This implies that orders for a customer cannot be entered in the database until the customer herself has been entered in the database (with an entry in table `Customer`). The rationale for this is that any order must come from a real customer, and as far as the database is concerned, the only real customers are those that are in the database. This makes sure that each order is connected to a real customer and there are no *dangling* references. In this sense, the database helps us keep our data internally consistent.

To help keep consistency, a database offers several (optional) commands when declaring foreign keys:

```
ON [DELETE|UPDATE] [CASCADE|SET NULL|SET DEFAULT|RESTRICT]
```

This instructs the database on what to do if the primary key being referenced is deleted or updated: we can also delete or update the foreign key that references it ('CASCADE'), set the foreign key to a NULL (foreign keys, unlike primary keys, can have null markers), set to a default value, or forbid the deletion or update of the primary key ('RESTRICT'). Suppose, for instance, that an existing customer has placed several orders. Then, there will exist one row in the `Customer` table with the information about this customer (and a `CustomerID`), and several rows in the `Orders` table, one for each order, and all of them using the customer ID to identify this customer. However, suppose that we made some mistake when entering the ID of the customer and we want to correct that (using UPDATE TABLE). The problem is that customer ID is used in `Orders` table as a value of a foreign key and would, after the change, not point to any existing customer. Clearly, the thing to do here is to propagate the change:

```
ON UPDATE CASCADE
```

Conversely, say the customer leaves for whatever reason and cancels her orders. When we delete the customer from the database (that is, from the `Customer` table), the customer ID is gone and, again, that `CustID` used in `Orders` does not refer to an existing customer anymore. In this situation, it may make sense to delete the order too, so we would use

```
ON DELETE CASCADE
```

Under other circumstances, another behavior may make more sense. We need to examine the semantics of the datasets and decide accordingly.

As this example shows, to represent a one-to-many relationship we can copy the key of the table representing the 'one' side to the table representing the 'many' side. However, many-to-many relationships need a different approach.

Many-to-Many Relationships

Suppose, as before, two datasets, one of chemical compound suppliers and one of laboratories, and a relationship between them where a lab buys some amount of a compound at some date. Then our database would look as follows:

```
CREATE TABLE SUPPLIER (
SupName character(100) PRIMARY KEY,
....);

CREATE TABLE LABORATORY (
LabName character(200) PRIMARY KEY,
....);

CREATE TABLE BUYS (
SupName character(100) FOREIGN KEY REFERENCES SUPPLIER,
LabName character(200) FOREIGN KEY REFERENCES LABORATORY,
CompoundId character(10),
Amount int,
Date date,
...
PRIMARY KEY (SupName, LabName));
```

As this example shows, in the case of a many-to-many relationship we create a separate table where we copy, as foreign keys, the primary keys of the tables representing the records involved in the relationships. These two foreign keys, together, are the primary key of this new table.

There are more complex cases, requiring further analysis. It is impossible to cover all of them and give hard and fast rules; database design is part art and part science. However, what has been shown in this section covers the most frequent situations and should help in all but a few cases.

Example: Very Complex Data

Assume that, as earlier, we have information about customers and their orders, connected by a one-to-many relationship. Further assume that we also have *point-of-contact (POC)* information for each customer, and that several customers have several POCs, although a POC relates only to one client. Thus, in addition to the tables shown in an earlier example, we also have

POC			
CustID	**POC-name**	**Phone**	**Email**
1	Road Runner	888-8888	rruner@gmail.com
1	ACME, Inc.	999-9999	manager@acme.com
2	Bugs Bunny	777-7777	bugsbunny@warnerbros.com
2	Duffy Duck	666-6666	duffyduck@warnerbros.com

Note that here Cust-id is also a foreign key to Customer. What would happen if we insist on having all information about our customers in a single table? We would have something like the following:

Customer-Order-POC								
CID	**Name**	**Address**	**OID**	**Date**	**TotalAmt**	**POC-name**	**Phone**	...
1	Wile E. Coyote	999 Desert Rd.	300	9/12/2010	20,000	Road Runner	888-8888	...
1	Wile E. Coyote	999 Desert Rd.	300	9/12/2010	20,000	ACME, Inc.	999-9999	...
1	Wile E. Coyote	999 Desert Rd.	301	1/2/2011	15,000	Road Runner	888-8888	...
1	Wile E. Coyote	999 Desert Rd.	301	1/2/2011	15,000	ACME, Inc.	999-9999	...
2	Elmer Fudd	123 Main St.	302	2/5/2011	500	Bugs Bunny	777-7777	...
2	Elmer Fudd	123 Main St.	302	2/5/2011	500	Bugs Bunny	777-7777	...
2	Elmer Fudd	123 Main St.	303	6/5/2012	800	Duffy Duck	666-6666	...
2	Elmer Fudd	123 Main St.	303	6/5/2012	800	Duffy Duck	666-6666	...

Note that in this table:

- Customers are repeated many times: a customer with n orders and m POCs will appear $n \times m$ times.
- on each customer, all combinations of Order and POC are shown. That is because Orders and POCs are *orthogonal*, that is, which orders a customer places do not depend on a particular POC, and who is a POC for a particular customer does not depend on that customer's (current) orders.

As a result, all data (that about customers, about orders, and about POCs) is repeated in the above table. The result can be pretty devastating for analysis that only wants to look at some of the data; an example of the problems that come up in this scenario is given in Sect. 5.1.

We close this section with one more consideration that is likely to appear in many real-life projects: time.

Example: Time in Databases

Often, databases record events that happen over time, hence temporal information is very important. Temporal information can make the schema of a database more complex. For example, in one of the examples above we described a clinical trial where participants had several vital signs checked daily. Note that a person has only one blood pressure *at a given moment*. Hence, when considering the person alone, the attribute is single-valued; however, when considering the person and the time of the measurement, it becomes a multi-valued attribute. Hence, we need to adjust our database accordingly. Assuming that all vital signs are measured together (at the same time), we could have tables

```
CREATE TABLE PARTICIPANT (
PId INT PRIMARY KEY,
....);

CREATE TABLE MEASUREMENTS (
PId  int FOREIGN KEY REFERENCES PARTICIPANT,
Date date,
Time time,
blood_pressure_sys int,
blood_pressure_dia int,
respiratory_rate int,
heart_rate int,
temperature float,
PRIMARY KEY (PId, Date, Time));
```

Note that we need to know the patient and the date and time to determine which vital signs we are talking about (if they were taken only once a day, patient and date would suffice).

We can now describe a database as a collection of tables where all tables are related—that is, each table has one (or several) foreign keys referencing other tables in the database or is referenced by other tables in the database (or both). The *schema of a database* is simply the collection of schemas of the tables in the database, including the foreign keys that link some tables to others. It is clear now that one of the reasons to give each table a primary key is so that it can be referenced by other tables in the database if need be.

Note that most approaches to data analysis require only relatively simple, tabular or near-tabular datasets that can be handled in a single table. However, it is important to be aware of the fact that databases can handle more complex cases whenever, for any of the reasons mentioned, data cannot fit in a single table. It is also important to know that data on a single table can always be normalized by splitting it into several, and that data in several tables can be combined to create a single table—both of those transformations are easy in SQL, as we will see.

2.3 Other Types of Data

Whether data fits into one table or several, it is still what we called *structured data* in Sect. 1.2. Relational databases were designed to deal with this type of data. However, over the years they have evolved to deal with semistructured and unstructured data as well. Here we discuss how to store such data in the database.

2.3.1 XML and JSON Data

As discussed in Sect. 1.2, XML and JSON are used with *hierarchical* or *tree structured* data. One way to see this type of data is as a chain of one-to-many relationships. What distinguishes this from the troublesome example we saw earlier is that the relationships are ‘chained’ together, with entities that are on the ‘one’ side of a relationship being on the ‘many’ side of another—hence forming a tree. Imagine, for instance, a university that is divided into various schools; each school is divided into several departments, and each department into several sections. Then we can think of the relation university–schools as one-to-many: the university is associated with several schools, but each school is associated with one university (in this case, there is only one university, but the pattern repeats at each level: a school is associated with several departments, but each department is associated with only one school. The same holds for departments and sections).

Hierarchical data can be stored inside the database in one of the two ways: by *flattening* it or by using the new XML or JSON data types added to the SQL standard.

In this first case, we can put hierarchies in tables in one of the two ways: normalized and unnormalized. Normalized data breaks the hierarchy into levels by representing each one-to-many relation in the hierarchy in its own table. Unnormalized data puts all the data in a single table by repeating data in one level as many times as needed to fill lower levels.

Example: Hierarchical Data

The hierarchy in Fig. 2.3 can be expressed in XML as

```
<superfamily name="Hominoidea>
  <family name="Hominidea">
     <subfamily name="Hominae">
       <tribe name="Homonini">
         <genera name="Hominini">
           <species name="Human"/>
         </genera>
         <genera name="Pan">
           <species name="Bonobo">
           <species name="Chimpanzee">
         </genera>
     </subfamily>
     <subfamily name="Ponginae">
         <genera name="Pongo">
           <species name="Orangutan"/>
         </genera>
     </subfamily>
    </family>
    <family name="Hylobatidae">
         <genera name="Hylobates">
           <species name="Orangutan"/>
         </genera>
    </family>
```

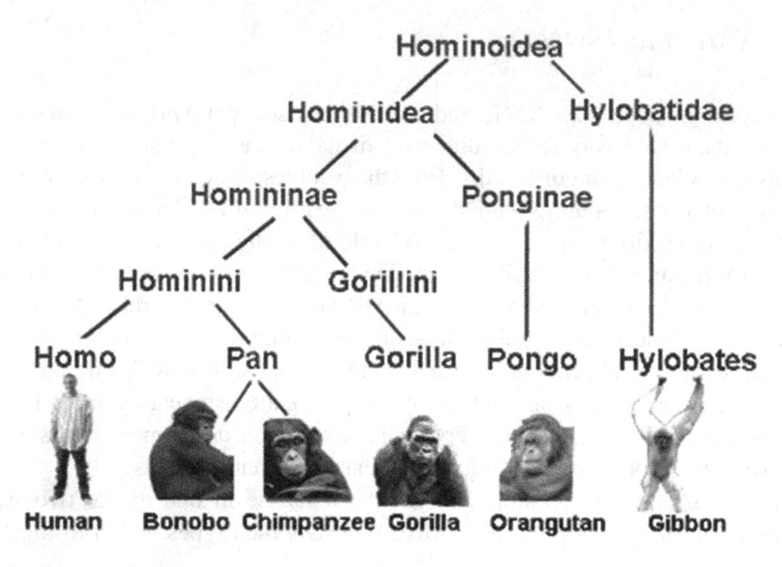

Fig. 2.3 An example of hierarchical data: biological classification

```
</superfamily>
```

In a flattened table, this becomes

Superfamily	Family	Subfamily	Tribe	Genera	Species
Hominoidea	Hominidea	Homininae	Hominini	Homo	Human
Hominoidea	Hominidea	Homininae	Hominini	Pan	Bobobo
Hominoidea	Hominidea	Homininae	Hominini	Pan	Chimpanzee
Hominoidea	Hominidea	Homininae	Gorillini	Gorilla	Gorilla
Hominoidea	Hominidea	Ponginae	Null	Pongo	Orangutan
Hominoidea	Hylobatidae	Null	Null	Hylobates	Gibbon

The schema has one attribute per hierarchy level, and the extension has one row per leaf value: values in higher levels are then repeated as necessary to fill in all rows (another way to see this is that each row represents the path from a leaf to the root). This causes redundancy, so we can split this table into a collection of tables:

Superfamily	Family
Hominoidea	Hominidea
Hominoidea	Hylobatidae

Family	Subfamily
Hominidea	Homininae
Hominidea	Ponginae

Subfamily	Tribe
Homininae	Hominini
Homininae	Gorillini

Tribe	Genera
Hominini	Homo
Hominini	Pan

Genera	Species
Homo	Human
Pan	Bobono
Pan	Chimpanzee
Pongo	Orangutan
Hylobates	Gibbon

Each table here represents the links connecting a pair of levels. This avoids repeating higher levels. Note, however, that if there are 'holes' in the hierarchy (as is the case here), reconstructing the original table from these smaller ones is, in general, *not* possible. For that, it is necessary to 'fill in' the holes with some artificial value, one different value per hole.

It is important to point out something about this example. We have seen before a situation with two one-to-many relationships, the example of customers, orders, and POC. There, there was a one-to-many relationship between Customer and Order (with Customer being the 'many' side) and another one-to-many relationship between Customer and POC (with Customer again being the 'many' side). However, orders and POCs were independent of each other. We called this case problematic, a situation that should be avoided.[9] However, in scenarios like the one above (with hierarchical data), we also have several one-to-many relationships, but no such problems. The crucial difference is in the "orientation" of the relationships. Suppose that instead of Customers, Orders, and POCs we have Customers, Orders, and Regions, where a customer is associated with one (and only one) region, but a region may have several customers associated with it. We again have two one-to-many relationships, one between Regions and Customers, and another one between Customers and Orders. However, now Customers play different roles in each relationship: it is the 'one' side with Regions and the 'many' side with Orders. This can be handled just like the examples shown in this section.

Another way to store the data above, especially if it comes with other (simple) data, is to create a table that has an XML or JSON attribute. In modern relational systems, a column can be declared as an XML or JSON type. This is a data type offered by the database system, so that such columns can contain data expressed in XML or JSON directly, without any need to flatten it.

Example: Hierarchical Data in XML/JSON

Using an XML type, the example above could be stored as follows:

```
CREATE TABLE HIERARCHY(
 ID INT PRIMARY KEY,
 Name character(50),
 Data XML);

INSERT INTO HIERARCHY VALUES(1, "Biological",
 "<superfamily name="Hominoidea">
    <family name="Hominidea">
       <subfamily name="Homininae">
         <tribe name="Homonini">
           <genera name="Hominini">
            <species name="Human"/>
```

[9]This is technically known as a *4th Normal Form* problem.

```
    </genera>
      ...
  </superfamily>");
```

Note that the XML value has been entered as a string, in quotes. A similar idea would work with JSON.

We will see later (Sect. 4.4) that it is possible to change data in XML/JSON to a flattened table using database functions, so that a large XML/JSON dataset can be loaded into the database directly as such and then transformed into a tabular format for analysis.[10] As stated earlier, most data mining and machine learning tools assume tabular data, so when presented with data in XML/JSON, one possibility is to load it into the database as such, using a table with an XML/JSON column, and then flattening this into a regular table, which can then be fed to the right tool.

2.3.2 Graph Data

Representing graph data in a relational database is very easy (doing interesting things with it is not so easy, as we will see in Sect. 4.6). There are, in fact, several options for dealing with graph data, but the most popular is to arrange the graph into two tables:

- A *nodes* table, where each node/vertex is represented by a row. The attributes that all nodes have in common constitute the schema of this table. If no primary key exists, some kind of identifier is generated and added to this table. For many real-life datasets, this works well since in many graphs nodes tend to represent the same type of entity and hence they are pretty homogeneous. In graphs where nodes are of several types, different tables may have to be used (one table per type). For instance, a graph that linked People and Books through a set of 'read' edges could use two separate tables, `People` and `Book`, to store the nodes.
- An *edges* table, where the links between nodes are stored. Each row in this table represents an edge, and the schema has, at the least, two attributes, one to represent each node involved in the edge (additional attributes can be used for graphs where edges are labeled or additional information is present). When the edges are directed, one attribute represents the *source* (or *origin*) nodes of each edge and another attribute the *destination* nodes, and edges are always interpreted as going from source to destination. When the edges are not directed, no distinction is made between node attributes. If it is necessary to make clear that the graph is undirected, we can either repeat each edge twice (once as (a,b),

[10]The opposite (data in a flattened table transformed into XML/JSON format) is also possible, but we do not cover it in this textbook.

another as (b,a)) or we can, in our queries, read the rows in either order (see Sect. 4.6 for examples).

Using generic SQL, the tables look like this:

```
CREATE TABLE nodes (
 id INTEGER PRIMARY KEY,
 name character(16) NOT NULL,
 feature1 datatype1,
 feature2 datatype2,
 ...);

CREATE TABLE edges (
 a INTEGER NOT NULL REFERENCES nodes(id)
            ON UPDATE CASCADE ON DELETE CASCADE,
 b INTEGER NOT NULL REFERENCES nodes(id)
            ON UPDATE CASCADE ON DELETE CASCADE,
 label character(256),
 PRIMARY KEY (a, b));
```

Example: Graph Data as Tables

The example of (fake) Tweeter users can be expressed in tables as follows:

```
INSERT INTO NODES VALUES
   (1, ''Shaggy'', ...), (2, ''Fred'', ...),
   (3, ''Daphne'',...), (4, ''Velma'', ...);

INSERT INTO EDGES VALUES
   (1,2, ''follows''), (1,3,''follows''),
   (1,4,''follows''), (3,4,''re-tweets''), (2,3,''likes'');
```

Another approach is to store graphs as matrices. For readers unfamiliar with the concept, one can think of a matrix as a representation of a set of values in a rectangle, determined by rows and columns. For instance,

$$\begin{bmatrix} a_{11} & a_{12} & a_{13} \\ a_{21} & a_{22} & a_{23} \end{bmatrix}$$

is a matrix with 2 rows and 3 columns representing 6 data values, $a_{11}, a_{12}, a_{13}, a_{21}, a_{22}$, and a_{23}. Each value a_{ij} sits in row i, column j.

The *adjacency matrix* of a graph with n vertices is an $n \times n$ matrix where each row/column represents a vertex in the graph (vertices in the graph are numbered $1, \ldots, n$ to facilitate this representation), and entry (i, j) in the matrix (the entry in row i, column j) represents information about the edge between nodes i and j, if such an edge exists. A *Boolean* adjacency matrix simply has a 1 to indicate that the edge exists and a 0 to indicate that it does not exist. In the case of matrices

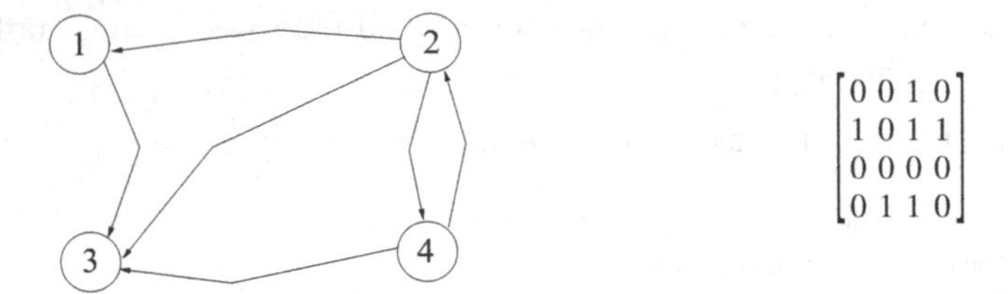

Fig. 2.4 Simple graph and its Boolean adjacency matrix representation

with weights in the edges, we get a *edge weight* matrix by setting entry (i, j) to the weight of the edge between i and j if such an edge exists, 0 otherwise. As an example, we show in Fig. 2.4 a very simple graph and its Boolean adjacency matrix. The graph has 4 nodes (nodes are numbered to make the representation clear) so the matrix is 4×4 (4 rows and 4 columns); by convention, rows are numbered top to bottom (so top row is row number 1) and columns are numbered left to right (so rightmost column is column 1).

The matrix M can then be represented via a table with schema *(row,column, value)*, where *row* and *column* are the integers giving the row and column of the matrix and *value* giving the appropriate entry. That is, if $M(i, j) = v$, we add row (i, j, v) to the table (it is a convenient convention not to add a row whenever $M(i, j) = 0$. This helps save lots of space, especially for *sparse* graphs, those with a low number of edges). Note that this is essentially the same as the EDGE relation described above; however, matrix algorithms treat this data differently. As we will see in Sect. 4.6, SQL can do matrix multiplication, sum, difference, multiplication by scalar, and transposition but cannot do more complex stuff (like finding linear independence (rank), determinants, eigenvector, and eigenvalues). Still, quite a few algorithms require nothing more complex than matrix multiplication, so this representation may be useful in some cases.

The matrix representation of a graph has some interesting properties:

- if there are no self-loops in the graph (no edges between a node and itself), the *diagonal* of the matrix (the collection of all the entries (i, i), that is, in row i and column i, for $i = 1, \ldots, n$) is all zeroes.
- if the graph is undirected, the adjacency matrix is *symmetric*, that is, entry (i, j) is the same as entry (j, i). This is not the case for a directed graph, where we store an edge only in one 'direction.'
- the most interesting property (for our present purposes) of adjacency matrices is that by multiplying the matrix by itself we get information about paths in the graph.[11] Let M be an adjacency matrix for graph G; then

[11]For readers not familiar with matrix multiplication, we explain the operation (and SQL code for it) in Sect. 4.6.

- $M^2 = M \times M$ represents paths of length 2 in G, that is, pairs of edges where the first edge goes from a node i to a node j, and the second edge goes from node j to a node k—this is a path of length 2 between i and k. If M is Boolean, then $M^2(i, j) = 1$ if and only if there is a path from node i to node j of length 2. If M is weighted, then $M^2(i, j)$ is the weight of the path of length 2 between i and j (if one exists; 0, otherwise).
- The above is true for any lengths, not just 2. In general, M^l (the product of M with itself l times) contains information about paths of length l. In particular, M^n is the adjacency matrix of the transitive closure of M (recall that n is the number of nodes in G, so there cannot be paths longer than n in G if we do not admit loops). Thus, we can use matrix multiplication to find out the existence of paths (using a Boolean matrix) or their length (using a weight matrix), even shortest paths.

While it is not convenient or possible to always use SQL for graph algorithms, there are some basic graph analysis tasks that can be carried out in the database. If nothing else, the database can be used to store and maintain very large graphs, which can later be downloaded (perhaps in small chunks) for analysis with graph tools, which tend to not do well with graphs that are larger than what a computer can handle in memory.

2.3.3 Text

Text refers to data expressed as fragments of natural language (English, Chinese, etc.). The unit of text is usually called a *document* and it may refer to something very short (a tweet, an email body) or long (a whole book). As stated earlier, usually text comes as a collection of documents, called a *corpus*; in some cases, documents in a corpus come with some structured data. For instance, a collection of emails may have attributes like *sender*, *receiver*, *date*, *subject*, and *body*, where *subject* is a short string and *body* is a string of arbitrary size, which can be considered text.

Text can be stored as an attribute of a table by taking advantage of a data type available in most systems: the *text* type. Essentially, this is a very long string, but in most systems it comes with functions that facilitate manipulation and analysis of strings (text analysis is described in Sect. 2.3.3). In Postgres, *text* is a variable, unlimited length string. In MySQL, there are types TINYTEXT (up to 256 characters), TEXT (up to 64,000 characters, approximately), MEDIUMTEXT (up to 16 million characters, approximately), and LONGTEXT (up to 4 billion characters, approximately). Even if the data is simply a corpus with no attached structured data of any type, it is possible to create a table with a single attribute in the schema, of type text, and to store each document in a row. This will enable basic analysis of the text like *keyword search* or *sentiment analysis* (see Sect. 2.3.3 for details).

Example: Text Data

In the case of the Hate Crime dataset from ProPublica, we had schema *Article-Date, Article-Title, Organization, City, State, URL, Keywords, and Summary*. As we saw, the last two attributes could be quite long and probably need to be analyzed in terms of their contents, so it makes sense to make them of type text. In Postgres:

```
CREATE TABLE HATE-CRIME(
article-date Date,
article-title char(256),
Organization char(128),
City char(64),
State char(2),
URL char(32),
Keywords Text,
Summary Text);
```

An insertion into this table looks as follows:[12]

```
INSERT INTO HATE-CRIME VALUES(
"3/24/17",
"Kentucky Becomes Second State to Add Police to Hate Crimes
Law",
"Reason",
"Washington",
"District of Columbia",
"http://reason.com/blog/2017/03/24/kentucky-becomes-second-
stateto-add-pol",
"add black blue ciaramella crime delatoba donald gay hate
 law laws lives louisiana matter police trump add black
 blue ciaramella crime delatoba donald gay hate law laws
 lives louisiana matter police trump",
"Technically, this is supposed to mean that if somebody
 intentionally targets a person for a crime because they are
 police officers, he or she may face enhanced sentences for
 a conviction. That is how hate crime laws are used in cases
 when a criminal targets");
```

Exercise 2.4 In the example of email collection, the data can be stored in a table with schema (email-id, sender, receiver, date, subject, body). Create a table in Postgres or MySQL for this dataset using text type for the last two attributes and try to insert some Enron data (or made-up data) into this table.

[12]We show each value in a separate line for clarity; it is not necessary to enter values this way. Putting the values in quotes and separating them by commas is what really matters.

2.4 Getting Data In and Out of the Database

Once a table or tables have been created in a database, it is time to bring in the data. We can also, once we have the data, make changes to it, delete (some subset of) it, or add more whenever additional data is available. The following subsections describe these activities.

2.4.1 Importing and Loading Data

There are several ways to do put data in the tables of a database, depending on the circumstances, but the two most common ones are to insert tuples or to load in bulk. The first one (which we have already seen) uses the `INSERT` SQL statement. This statement allows us to add one row (or a few rows) to an existing table, and it can be used whenever we need to add data in small quantities and we want to have total control on how data is entered in the database. However, this statement is too tedious and error-prone when we already have a non-small dataset that needs to be added to the database.

If data is already in a file, it is possible to *load* the data into the database in one swoop. All database systems have some command which takes a file name and a table name and brings in the data from the file into the table—as far as the data in the file is compatible with the schema of the table. This command assumes that the data in the file can be broken down into lines, each line corresponding to a row/tuple for the table.

The *load* command is system dependent, although the basic outline is the same for most systems. Generally, one specifies the location of the file in the computer, the table to load into, and provides a description of how the lines in the file are to be broken down into the values that make up a row/tuple by indicating how values are separated from each other and a few other characteristics. For instance, the typical `csv` files use the comma as a separator; other typical separators are the semicolon character or the tab. The table must already exist before this command is used, and the data in the file must fit into the schema of this table. The loader will read the file line by line and split each line according to the given instructions. It will expect that there are as many values on each line as there are on the schema of the table and of the right type. When this is the case, the loader will create a tuple/row for each line in the file and will insert it in the table.

However, datasets often come with *dirty data* and this creates a problem when trying to load the data in the database. Some of the most common problems are:

- missing values. Lack of values can be manifested in a file in two ways: by an *empty field* (that is, two consecutive appearances of the delimiter) or by some marker (markers like '\N' or 'NA,' for 'Not Available,' are particularly common). Dealing with empty fields is relatively straightforward. Some systems will automatically set the corresponding attribute in the table to a *default* value:

the empty string, for string types; zero, for numeric types; and the date or time 'zero' for dates and times. Others will create a NULL marker in the database. However, dealing with markers tends to be messy. The first problem is that different datasets may use different conventions to mark missing values; the dataset may need to be explored *before* it can be loaded in the database to identify such markers.[13] The second problem is that null markers may confuse the loader about the type of data it is reading. In numerical attributes, the system expects strings that can be transformed into numbers (essentially, strings made up of digits and optionally a hyphen (-) or a period (.). Even numbers with commas can create problems). When a string like 'NA' is found, the system is unable to transform it into a number or recognize it as a missing value marker. The same problem happens with temporal information, where the system expects a string in a certain format that it can parse and recognize as a date or time. Even in string based values, the system may likewise confuse a string like 'NA' with a valid value, not a missing value marker. The best way to avoid errors is system dependent: in some systems, it is better to delete unrecognized markers and leave empty fields; in others, it may be necessary to create a special value of the right type (for instance, using -1 for a numeric field that contains only positive values).
- strings are represented differently in different datasets. In some cases, strings are stored by surrounding them with single quotes ('), sometimes with double quotes ("); sometimes they are stored without any quotes. This can cause confusion in the loader, especially when strings include characters other than letters or numbers. For instance, the string "Spring, Summer, Fall, Winter ... and Spring" is the title of a movie, but it includes 3 commas and 3 dots and may confuse a loader when stored in a CSV file—the commas on it may be confused with separators.[14] Again, solutions depend on the system: some systems are smart enough to leave everything in quotes alone (in which case, making sure that all strings are surrounded by quotes is the way to go); others may require that those extra commas go away.
- dates and times are typically recognized by most database systems if they follow a certain format (we discuss such formats in detail in Sect. 3.3.1.3). When the data in the file does not conform to the format, the system tends to read it as a string.

Different systems use different tricks to help deal with these problems. In MySQL, one would write

```
LOAD DATA INFILE filename
INTO TABLE table
 [FIELDS [TERMINATED BY string]
         [ENCLOSED BY char]
         [ESCAPED BY char]]
```

[13]Command line tools are the appropriate tools for this task [9].

[14]This example is from a dataset in the Imdb website.

in order to load data from file `filename` into table `table`. The optional TERMINATED BY clause allows the user to specify the character that separates one field from the next; the default is tab, but it can be changed to comma or another character with this clause. The optional ENCLOSED BY clause allows the user to indicate how values of type string are enclosed; usually, this is a quote or a double quote, although other values can be used. Finally, the optional ESCAPED BY is used to indicate how to handle special characters: these are usually indicated by a backslash ('\') followed by a character. For instance, '\n' denotes the newline (linefeed) character, a control character used to separate lines. In addition, a statement `IGNORE n LINES`, where *n* is a positive integer, can be used if we need to not load the first *n* lines of the file; in particular, `IGNORE 1 LINES` is used when the data file contains a *header*, which we do not want to load (it could be confused with data).

Exercise 2.5 Load data from the file ny-flights.csv into the table `ny-flights` in MySQL. Caution: this can be a long process, depending on the computer. As an alternative (and to be able to fix errors easily), start with a small sample by choosing the first 1000 lines or so.

In MySQL, it is possible to perform some data transformations on the LOAD command by using *user variables*. User variables are labels that start with '@'; they can be used in assignment operations, as the following example shows.

Example: LOAD and Data Transformation in MySQL

The following example assumes a file T with schema `(column1,column2)`; it uses the first input column directly for the value of `T.column1` and divides the value of the second column in the file by 100 before using for `T.column2`:

```
LOAD DATA INFILE 'file.txt'
  INTO TABLE T
  (column1, @var1)
  SET column2 = @var1/100;
```

This approach can be used to supply values not derived from the input file. The following statement sets `T.column3` to the current date:

```
LOAD DATA INFILE 'file.txt'
  INTO TABLE T
  (column1, column2)
  SET column3 = CURRENT_DATE;
```

You can also discard an input value by assigning it to a user variable and not assigning the variable to a table column:

```
LOAD DATA INFILE 'file.txt'
  INTO TABLE T
  (column1, @dummy, column2, @dummy, column3);
```

This statement reads and uses the first, third, and fifth columns of the file and reads but ignores the second and fourth columns.

When the LOAD DATA statement finishes, it returns a message indicating how many records (lines) were loaded, how many were skipped (because of some problem), and how many warnings the LOAD process generated. This can be used to check whether there were problems and, if so, how many.

In Postgres, the equivalent command is called COPY. COPY FROM copies data from a file to a table (appending the data to whatever is in the table already). The format is

```
COPY table_name [ ( column\_name [, ...] ) ]
 FROM  'filename'
 [ [ WITH ] ( option [, ...] ) ]
```

The optional WITH option is used to give different hints as to how to handle the data. The most important option is FORMAT, which determined the data format to be read or written: TEXT (default), CSV, or BINARY. Another useful option is NULL AS, which is used to specify which characters or strings should be interpreted as NULLS; this is very useful when the dataset uses its own convention to mark missing values (the default in CSV is an empty value with no quotes, that is, two commas together or a comma at the end of the line). Finally, QUOTE AS can be used to tell the system how strings are quoted (single or double quotes; the default in CSV mode is double).

When the TEXT format is used, COPY FROM will raise an error if any line of the input file contains more or fewer columns than are expected. Hence, it is important to make sure that the right delimiter and the right options are used, so that each line can be parsed correctly.

Option HEADER specifies that the file contains a header line with the names of each column in the file. This option is allowed only when using CSV format.

Example: Loading Data in Postgres

To copy data from a csv file containing a header, we would use a command like the following:

```
COPY ny-flights FROM '~/DATA-MNGMNT/DBS/ny-flights.csv'
 DELIMITER ',' WITH FORMAT CSV HEADER;
```

Note that the DELIMITER specification is redundant in this case.

Exercise 2.6 Load data from the file ny-flights.csv into the table `ny-flights` in Postgres.

2.4.2 Updating Data

Once data is in the table, we may need to modify some of it. Modifications are accomplished with two SQL commands: DELETE and UPDATE.

The DELETE command is used to delete existing data. The format of this command is

DELETE FROM table-name WHERE condition

Here condition is an expression that is evaluated in each row in table-name and results in a True or False value. In its simplest form, it involves an attribute name from the schema of the table, a comparison operator (like '=,' '≤,' <) and a constant. For instance, in table Chicago-employee, a condition could be: name = 'Jones'. In each row of the table, the system will look up the value of attribute name and see whether this value is 'Jones' (in which case the condition evaluates to True) or something different (in which case the condition evaluates to False). More complex conditions can be built from simple ones (conditions are explained in depth in Sect. 3.1). The DELETE statement works as follows: on each row of the table, the condition is checked. If it returns True, then the row is deleted. If the condition returns False, the row is left unchanged on the table. Clearly, the condition controls the effect of this statement, and it must be written carefully. In particular,

- if the condition is not true in any row of the table, the command has no effect, as the table is left unchanged.
- if the condition is true in every row of the table, all the data in the table is erased (this will also happen if no condition is given, see below). The table is now empty.

It is customary, when using this command, to use a primary key or a foreign key in the condition, in order to control exactly where the command is applied.

Example: Table Deletion

Suppose that in table ny-flights we are told that the flight from EWR to IAH on January 1st, 2013 was canceled. We could use this information directly in SQL to update the data:

```
DELETE FROM NY-FLIGHTS
WHERE year = 2013 and month = 1 and
      day = 1 and origin = 'EWR' and dest = 'IAH';
```

But this will delete any tuple that fulfills the conditions—that is, if there are several flights on that day from EWR to IAH, all of them will be deleted. A better idea is to find out the flight id of the canceled flight uses a condition like id = ... to make sure we are singling out for deletion only the right flight. In general, using a key in the condition (primary or unique) is the safe way to proceed.

Exercise 2.7 Delete all flights that go from JFK to ATL in the afternoon (departure time between noon and 5 pm).

The DELETE command can be issued *without* a WHERE clause, in which case it will delete all the data in the table (in other words, in the absence of a condition, the command applies to every row). Note the difference with the DROP TABLE command: the DELETE will get rid of the data, but leave the table (empty) in place;

the DROP TABLE will get rid of table and data. After a `DELETE`, we can continue using the table (for instance, putting some data on it again); after the `DROP TABLE`, the table itself is gone so if needed, it would have to be recreated (with a `CREATE TABLE` statement).

The `UPDATE` command is used to change existing data. By 'updating' a row we mean to change the value of one or more of the attributes for that row (this is called 'modifying' the row in some textbooks). The format of this command is `UPDATE table-name SET attribute = value WHERE condition`

Here, `condition` is exactly the same as in the DELETE statement; changes are made only to tuples on which the condition is true. In addition, the SET clause tells the system exactly what changes to make: `attribute` is the name of an attribute in the schema of the table, `value` is a value of the right data type. It is possible to make several changes at once by giving a sequence of `attribute = value` expressions separated by commas. Any attributes not mentioned are left unchanged.

Example: Table Update

Suppose that in table `ny-flights` we are told that the flight with id 1 did have flight number 8501, not 1545 as it appears in the data. We can change the data with

```
UPDATE NY-FLIGHTS SET flight = "8501" WHERE id = 1;
```

As in the case of deletion, we can use any condition, but it is best to use a condition involving a primary key to make sure we only change the row (or rows) that we want to change.

It is important to make clear that the old data is gone forever. If we want to keep old data and just register that there has been a change, we can create *versions* of data. For this, a temporal attribute is added to the table to reflect when data is valid; instead of updating existing values, new rows are inserted with the new value, and old ones are left in place—the temporal attribute is used to tell which row reflects the *current* situation and which ones are *historical* data. This may be important in applications that want to examine changes over time, and it is the foundation of *time series* data.

Note that one could change a row by first deleting it and then inserting it with the new values. However, doing an update is almost always preferred, as it is much more efficient.

Exercise 2.8 Change the destination of all flights that fly into JFK to LGA.

2.4.3 Exporting Data

Sometimes we are interested in taking data *from* the database and putting it in a file, so we can use it with other tools. This is called *exporting* or *dumping* the data. We also want, from time to time, to make a copy of data in the database in case there is a serious problem, to avoid data loss.

The process for saving the data into a file for *backup* purposes is also, like the load, system dependent, although the idea is basically the same for most databases. In MySQL, there are two basic procedures for exporting data. The first one is to use the reverse of the LOAD statement:

```
SELECT columns INTO OUTFILE filename
 [FIELDS [TERMINATED BY string]
         [ENCLOSED BY char]
         [ESCAPED BY char]]
FROM table-name;
```

This command will take the data in table `table-name` and put it in the file `filename`. If the file already exists, this statement gives an error; the user should specify a new file name for this command. As we will see later, this command can be used to extract only certain parts of a table or database into a file. This is useful when we want to carry out focused analysis (using R or other tools) on part(s) of a large database.

The other way to get data out of a MySQL database is to execute a `mysqldump` command. This command generates an SQL script, that is, a file with SQL commands; it is customary to give such files a `.sql` extension. The script contains step-by-step instructions (in SQL) to recreate a table or a database. For a single database, the command is written (in the command line):

```
mysqldump --databases database-name > mydump.sql
```

or as

```
mysqldump database-name > mydump.sql
```

Both commands instruct the system to save the script as a data file called `mydump.sql`. The difference between the two preceding commands is that without `-databases`, the dump output contains no CREATE DATABASE or USE statements.

To dump only specific tables from a database, name them on the command line following the database name:

```
mysqldump database-name table-name > mydump.sql
```

To reload a dump file written by mysqldump that consists of SQL statements,

```
mysql < mydump.sql
```

Alternatively, from within MySQL, use a source command:

```
mysql> source dump.sql
```

Both commands execute all the statements inside the SQL script `mydump.sql`, recreating whatever data was on the database or file we dumped.

In Postgres, one can also reverse the COPY FROM command using COPY TO:

```
COPY { table_name [ ( column_name [, ...] ) ] | ( query ) }
    TO  'filename'
    [ [ WITH ] ( option [, ...] ) ]
```

Besides getting data from an existing table (using `table_name`), one can pick certain data from the database using the 'query' option. A *query* is the name for a process to pick certain data in the database; it is a fundamental tool in data analysis, and it is explained in the next chapter.

In the options, DELIMITER specifies the character that separates columns within each row (line) of the file (the default is a tab character in text format, a comma in CSV format); HEADER specifies that the schema of the table is written to the first line of the file. The filename must be specified as an absolute path. As in the case of MySQL, this command can be used to extract only certain parts of a table or database.

The command equivalent of `mysqldump` in Postgres is called `pg_dump`. To dump a single table into a file, the syntax is

```
$ pg_dump -t table-name mydb > filename.sql
```

As before, this creates a SQL script. To get the data back from the file into the table, use the command

```
$ psql -d new-table-name -f filename.sql
```

A whole database can also be dumped into a file:

```
$ pg_dump database-name > filename.sql
```

As before, this SQL script can be used to recreate the database from scratch with:

```
$ psql -d new-database-name -f filename.sql
```

The command `pg_dump` can also be used to archive a database, by using the `-F switch`. For instance,

```
$ pg_dump -Fd database-name -f dumpdir
```

will create a directory-like archive of the given database. To restore this archive, the command `pg_restore` is used:

```
$ pg_restore -d new-database-name dumpdir
```

Exercise 2.9 In either Postgres or MySQL, export the table `ny-flights` into an SQL script. Drop the table and restore it using the saved script.

Chapter 3
Data Cleaning and Pre-processing

In this chapter, we introduce the basic tool that SQL uses to extract information from a database, that is, the SELECT statement. We then show how to use it to carry out some basic tasks to be done during the Exploratory Data Analysis (EDA), data cleaning, and pre-processing stage of the data life cycle (see Sect. 1.1). After introducing the basic blocks of SELECT in the next section, we discuss EDA in Sect. 3.2, data cleaning in Sect. 3.3, data pre-processing in Sect. 3.4, and the implementation of *workflows* in Sect. 3.5.

3.1 The Basic SQL Query

SQL offers a statement, called the SELECT statement because of its initial keyword, in order to ask questions (in database parlance, *queries*) of the data. This statement allows us to extract information from the database, but it is also used when manipulating data. It is the most useful, used, and complex of all SQL statements. In this chapter, we show its basic structure; a discussion of additional features is left for Chap. 5.

In its basic form, the SELECT statement is written as

```
SELECT result-list
FROM data-sources
WHERE condition;
```

where

- `result-list` is a list of attribute names, or functions applied to attribute names (see Sect. 3.1.2). This list determines what gets retrieved from the database; hence, this list represents what information we are interested in getting as an answer to our question.

A. Badia, *SQL for Data Science*, Data-Centric Systems and Applications,
https://doi.org/10.1007/978-3-030-57592-2_3

- `data-sources` denotes where the data examined for this query comes from. There are basically two types of data sources in SQL:
  - Tables from the database: the FROM keyword is followed by one or more table names, separated by commas. The data in the extension of the table(s) is used for processing the query.
  - Another query: a whole query, i.e. an expression:
    `(SELECT .... FROM ... WHERE ...) AS new-name`
    can be used inside the FROM clause. This query is followed by the expression `AS new-name` because it has be given a name. The rationale for this is that, as we will see, all queries in SQL return a table as an answer. Hence, queries (including subqueries) can be seen as denoting a table (the answer to the query). When there is a query A (called a *subquery*) in the FROM clause of another query B (called the *outer* or *main* query), the system first evaluates A and uses the result of this evaluation (which is a table) as one of the data sources for evaluating B. The reason that subqueries use `AS table-name` to give their result a name is that this name is then used in the main query to refer to the result. Since B acts upon the results of A, using subqueries is a very useful tactic for computing complex results by breaking the process down into steps: the data is prepared with query A, and the final result is obtained with query B. We will see many examples where this approach is used.

  It is possible to mix, in the same FROM clause, table names and subqueries. This is because in the end, both table names and subqueries denote table extensions, i.e. collections of data. When more than one table is used, the extensions of all the tables are combined into a single table/extension. Exactly how this is done depends on the context and is discussed in depth in Sect. 3.1.1.
- `condition` is an expression that is evaluated on each row of the data extension, and it returns True or False. This is similar to the conditions used in the DELETE and UPDATE statements (see previous chapter). A simple condition compares the value of some attribute with a constant. More complex conditions can be formed by taking the conjunctions or disjunction of two conditions, or the negation of a condition.

Each query is evaluated as follows: the collection of rows denoted by the data sources in FROM are combined into a single table (if there are subqueries, these are processed first to obtain a result/table for each subquery); then the condition in WHERE is evaluated on each row. Rows where the condition is not True are disregarded, and those where the condition is True are kept (the WHERE clause may be absent from a query; in this case, all rows of the data source(s) are used). Using these filtered rows, the values of the attributes in `result-list` are picked up. Note that, for this evaluation to work, it must be the case that every attribute mentioned in the `condition` or in the `result-list` is present in the schema of the tables in FROM. If this is not the case, the system will not attempt to evaluate the query; it will return an error message instead of a result.[1]

Example: Simple Query

Looking at the data in table `ny-flights`, we may want to answer several questions. For instance, we may ask: are there any flights into JFK on November 10?

```
SELECT id
FROM NY-FLIGHTS
WHERE year = 2013 and month = 11 and day = 10
      and dest = "JFK";
```

In this example, the data source is the table `ny-flights`; the condition is a conjunction of 3 simple conditions, and only the attribute `id` is retrieved. This is an *existential* query (one asking if data fulfilling some constraints does exist), so retrieving the primary key is enough. The system evaluates this query as follows: the extension of table `ny-flights` (all the rows on it) is examined; on each row, the condition of the WHERE clause is evaluated (all 3 simple conditions are applied, and if all 3 return True, the whole condition returns True).[2] The collection of rows where the condition returned True is then processed; on each row, the system picks the desired attributes, as specified in the SELECT (in this case, `id`).

Example: Another Simple Query

Using again table `ny-flight`, we ask where those flights coming into JFK on December 10 are coming from, but only for trips longer than 1,000 miles. We can reuse our previous query as follows:

```
SELECT origin
FROM (SELECT id, distance, origin
      FROM ny-flights
      WHERE year = 2013 and month = 11 and
            day = 10 and dest = "JFK") AS T
WHERE distance > 1000;
```

In this example, the evaluation proceeds as follows: the subquery in the FROM clause is done first, starting with table NY-FLIGHTS and applying the WHERE conditions to it (so only flights from 11/10/2013 and destination 'JFK' are used) resulting in a (temporary) table named T (as indicated by the `AS T`)[3] with 3 columns

[1]This is typically the case when the user makes a typo or forgets the exact name of an attribute.

[2]Complex conditions in SQL are evaluated using the traditional rules of Boolean logic: for `A and B` to be true, both `A` and `B` must be true; for `A or B` to be true, it is enough that one of `A` or `B` is true (but it is okay if both are); for `not A` to be true, `A` must be false.

[3]In some systems, including MySQL and Postgres, giving an *alias* (a temporary name) to any table created by a subquery in FROM is required; in some, it is optional.

(attributes id, distance, origin); using this table, the main query is run: the condition distance $>$ 1000 is evaluated, and for all surviving rows, the origin attribute is picked. Note that in the subquery we selected attributes distance and origin because they were necessary in the following step. Note also that in this case, we could have written the query in one step, as follows:

```
SELECT origin
FROM ny-flights
WHERE year = 2013 and month = 11 and
      day = 10 and dest = "JFK" distance > 1000;
```

It will often be the case that there is more than one way to write an SQL query. Using subqueries is convenient in complex cases, as it breaks down a problem into simpler sub-steps. We will see other examples where using a subquery is a good idea.

Conditions depend on the type of the attribute(s) involved. There are specific conditions for numbers, strings, and dates. We introduce a few basic ones here, and more as we go along. For number, arithmetic conditions (comparisons using $<$, $\leq$, $=$, $>$, $\geq$) are common. For strings, it is possible to compare two strings with equality, but it is also possible to compare a string with a *string pattern*, an expression that requires the occurrence (or non-occurrence) of certain characters in the string in a certain order. The SQL standard requires the predicate LIKE, which takes a string and a simple pattern and compares them. Suppose, for instance, that we are unsure about an airport name, then a predicate like

dest LIKE 'SD_'

stipulates that the value of dest must start with 'SD' but has an additional character after that (the '_' stands for any character). The '*' symbol can also be used in patterns for LIKE; it stands for 'no character, or any one character, or any string of characters'). Most systems allow for much more complex patterns, including what is called a *regular expression*. We do not cover regular expressions in this book, but we discuss string functions in some depth in Sect. 3.3.1.2.

Another available predicate (both for numbers and strings) is IN. It compares an attribute to a list of constants; if the value of the attribute equals any of the constants, the IN predicate is satisfied.

Example: IN Predicate

We want to retrieve all flights that go into New York City. However, going to New York can be accomplished by flying into any of its two airports (JFK and LGA) and even by flying into neighbor New Jersey (Newark, EWR). Any one of them will do, so we write

```
SELECT *
FROM ny-flights
WHERE dest IN ("JFK", "LGA", "EWR");
```

Note that IN is equivalent to a disjunction: `attribute IN` $(value_1, \ldots, value_n)$ is semantically equivalent to

`attribute =` $value_1$ `or ...or attribute =` $value_n$[4]

so we could have written

```
SELECT *
FROM ny-flights
WHERE dest = "JFK" or dest = "LGA" or dest = "EWR";
```

Exercise 3.1 Select all the flights from `ny-flights` that fly on weekends (Saturdays or Sundays).

Finally, a predicate called BETWEEN is also offered. Like IN, it is really a shortcut. A predicate of the form `attribute BETWEEN` $value_1$ `and` $value_2$ is short for `attribute` $\geq$ $value_1$ `and attribute` $\leq$ $value_2$. This predicate requires an order in the attribute domain (i.e. the attribute must be an ordinal or a number); that is why it is commonly used with dates and numbers.

Example: BETWEEN Predicate

Suppose we are interested in the origin of all flights that landed between 4 and 5 am. The query

```
SELECT origin
FROM ny-flights
WHERE arr_time BETWEEN 400 and 500;
```

will retrieve all information about such flights.

Exercise 3.2 Select all the flights from `ny-flights` that fly longer than 500 but less than 1000 miles.

It is possible to refer, in the SELECT clause (and other clauses that we introduce shortly), to the attributes of a table by a number. This number refers to the position of the attribute in the table, as given when the table was created. That is, if the command:

```
CREATE TABLE foo(
A integer,
B varchar(48),
C float);
```

[4]This equivalence, like many others, only holds if there are no nulls involved. See the discussion of null markers in Sect. 3.3.2.

is used, attribute A can be referred to as 1, attribute B can be referred to as 2, and attribute C can be referred to as 3. The query

```
SELECT  1, 3
FROM foo;
```

is equivalent to

```
SELECT  A, C
FROM foo;
```

However, this practice is not recommended, as it can make queries difficult to read. The attributes retrieved in the SELECT clause become the schema of the result table, and it is a good idea to make sure we are retrieving exactly what we want. These attributes can be given a new name in the result, using the AS keyword:

```
SELECT dest AS destination, flight  AS flight-number
FROM ny-flights;
```

will produce a result with schema (destination, flight-number). This is usually done when saving the result to be used later or shown to others; we will see how to save the results of a query in Sect. 3.5. A query can also be used with the command to download data from the database into a file, so this mechanism can also be used to give our data meaningful names for data sharing.

What if we want *all* the attributes of a table? A real-life table may have dozens of attributes; listing them all makes writing queries long, tedious, and error-prone. SQL provides a shortcut: the '*' ('star' or 'asterisk') can be used in a SELECT clause to mean "all attributes."

Example: Getting All Attributes

The query

```
SELECT *
FROM ny-flights
WHERE dest = "JFK" and arr_time < 800;
```

will return all information (all attributes in the schema) about flights coming from "JFK" and arriving before 8 am.

One final thing to note about the results is that sometimes we may get *duplicates*, that is, two or more rows that contain the exact same values. This may happen because the table that we used in the FROM clause contained duplicates or because duplicates are created while processing the query.[5] When we want to get rid of duplicates, SQL provides the keyword DISTINCT that can be added to SELECT to signal to the system that we want duplicates eliminated.

[5]Recall that we said that the same tuple may be inserted multiple times in a table. But even if all rows in a table are different from each other, all this means is that two arbitrary rows are

Example: Eliminating Duplicates

The query

```
SELECT carrier, origin
FROM ny-flights;
```

will retrieve the airlines and the places that they fly from in our New York flight dataset. Note that the rows of the original table are likely all different, but there also very likely share values for some attributes. In this example, each row in the original table is a different flight; however, when focusing on carrier and city, we are going to see duplicates if the same carrier has more than one flight starting at a given city. To eliminate these duplicates, we could use

```
SELECT DISTINCT carrier, origin
FROM ny-flights;
```

The typical use for DISTINCT is to identify the different values that make up an attribute, especially a categorical/nominal one. When applied to a numerical attribute, DISTINCT will produce all unique values, and this will make a difference when applying arithmetic or other mathematical functions, as we will see.

Exercise 3.3 Make a list of all the carriers mentioned in the `ny-flights` dataset. That is, show all the unique contents of the `carrier` attribute.

3.1.1 Joins

As stated in the previous section, a FROM clause may mention more than one data source. If that is the case, the system combines all data sources into a single table. This is accomplished through one of two ways: *Cartesian products (henceforth, CPs)* and *joins*. Both play an important role in Data Analysis.

When the tables are simply listed in the `FROM` clause, the system will automatically produce the *Cartesian product* of all tables named and/or denoted by subqueries. The Cartesian product of two tables T and R yields a single table with a schema, which is the concatenation of the schemas of T and R (i.e. all attributes in T followed by all attributes in R); tuples for this table are created by combining tuples in T and tuples in R (all tuples in T are combined with all tuples in R).

---

different from each other *in at least one attribute*. Two rows may coincide in some attributes and be different in others. If our query retrieves only those attributes where the rows coincide, we will obtain duplicates.

Example: Cartesian Product

Explaining the behavior of CPs is best done through an (artificial) example. Assume tables T(A,B,C) and R(D,E), with these extensions:

T		
A	**B**	**C**
a_1	b_1	c_1
a_2	b_2	c_2
a_3	b_3	c_3

R	
D	**E**
d_1	e_1
d_2	e_2

In SQL, the CP is written as:

```
SELECT *
FROM T,S;
```

This produces a table with schema (A,B,C,D,E) and the following tuples:

CP of **T, R**				
A	**B**	**C**	**D**	**E**
a_1	b_1	c_1	d_1	e_1
a_1	b_1	c_1	d_2	e_2
a_2	b_2	c_2	d_1	e_1
a_2	b_2	c_2	d_2	e_2
a_3	b_3	c_3	d_1	e_1
a_3	b_3	c_3	d_2	e_2

Note that **T** has 3 tuples and R has 2 tuples, and their CP has $3 \times 2 = 6$ tuples. Each tuple in the CP is made up of a tuple from T (giving values to attributes A, B, and C) concatenated from a tuple from S (giving values to attributes D and E).

CPs can be used to combine more than two tables by combining them pair-wise: given 3 tables T, R, and S, we can take the CP of T and R and then the CP of the resulting table and S, to produce a single table. Because CPs are associative, we can do this in any order, and the result is still the same.

Example: A Common Cartesian Product

Sometimes when analyzing data, we need to compute a single value. In SQL, this will be done with a query. As stated earlier, all SQL queries return a table; what happens in this cases is that we have a table with a single attribute and a single row. For instance, a query may compute a single value v called F:

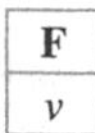

F
v

It is a common pattern to combine such a result with a data table. The Cartesian product of the above table R and table T from the previous example yields this result:

A	B	C	F
a_1	b_1	c_1	v
a_2	b_2	c_2	v
a_3	b_3	c_3	v

The net effect is to add the single value v to each row of T; now the value v can be compared to values in every row in the data. Note that the result has $3 \times 1 = 3$ rows. This is a common pattern that will be used over and over in what follows (usually, R is some statistic on the dataset itself).

In general, the size of the CP of T and R is the product of the size of T and the size of R, so this can get very large if T and R are even medium-sized tables: for instance, if T and R have 1,000 tuples each (a very small size for real-life datasets), their CP has 1 million rows (a more respectable size). CPs are only needed in certain occasions, as we will see; therefore, when writing an SQL query we always need to check, if we are using a CP, whether it is really needed. A typical use of CP is illustrated in the previous example; because one of the tables has size 1, there is no 'blow up' effect on the data size. However, beyond this pattern, CPs should be used with caution.

The other way of combining tables (and by far the more common) is to specify a *join* between them. To join two tables, it is necessary to give a condition involving attributes of both tables. This is written in SQL in one of two ways:

```
FROM Table1, Table2
WHERE attribute1 = attribute2

FROM Table1 JOIN Table2 on (attribute1 = attribute2)
```

Here, `attribute1` comes from (the schema of) `Table1`, and `attribute2` comes from (the schema of) `Table2`. What this condition does is to constrain which rows of `Table1` should be paired with rows of `Table2`. Among all tuples in the CP of the tables, only those that fulfill the condition are kept.[6] This allow us to combine data from two tables but to keep all the combinations that 'make sense.'

In order for the comparison to be possible, `attribute1` and `attribute2` should have the same data type. Also, for the result to be meaningful, both attributes should have related values. If you recall the discussion of Sect. 2.2, when information in two tables is related, we use foreign keys to indicate this fact. 99% of joins are on a primary key/foreign key connection—that is, `attribute1` is a primary key and `attribute2` is foreign key that refers to it, or vice versa. The reason is that,

[6]Technically, a join is a CP followed by a selection on the result of the CP.

when designing a database, we used the foreign key to connect related data across tables (see Sect. 2.2); therefore, it makes sense to put data back together using such connections. It is possible to use arbitrary conditions when writing a join; however, joining tables using something else than a primary key/foreign key connection rarely makes sense.

Example: Join

Recall the example where we had transaction data in a table and we split the table into two in order to avoid redundancy. We can now see how in SQL we can go from a single table to several, and vice versa. Joining the smaller, normalized tables `Transactions` and `Products` produces the table `Transaction-and-Products` with all the data combined:

```
SELECT *
FROM Transaction, Items
WHERE Transaction.transaction-id = Items.transaction-id;
```

Note that `transaction-id` is a primary key in `Transaction` and a foreign key in `Items`; hence, this is an example of a primary key/foreign key join.

Likewise, if we are given the full table `Transaction-and-Products` and we decide to split it into two normalized tables, this can be done by *projecting* (database parlance for picking only certain attributes from a table):

```
SELECT transaction-id, product
FROM Transactions-and-products;

SELECT distinct transaction-id, date, time, store
FROM Transactions-and-Products;
```

Note the use of `DISTINCT` on the case of `Transactions`; this is to combat the redundancy introduced by the design of `Transactions-and-products`.

Exercise 3.4 It is easy (and recommended) to check that the queries in the previous example indeed produce the expected results with the data in the example of the previous chapter.

The above example hints at an important technical detail: the schema of a CP or a join is created by concatenating the schemas of the relations being combined. We stated earlier that, in a given schema, each attribute name must be unique. However, when two schemas are combined, there is no guarantee that this will be the case. This is indeed what happens in the example above, where both table `Transaction` and `Product` have an attribute named `transaction-id`. The problem is that mentioning attribute `transaction-id` in the query now would create an ambiguity: there are two attributes with that name. To break this ambiguity, SQL allows the `dot notation`, `table-name.attribute-name`, where each attribute is preceded with the name of the table it belongs to, both names separated by a dot ('.'). This is exactly what we used in the example above and can always be used as far as each table has a different name—as they should inside a database.

Exercise 3.5 The `world` database is a sample database available from MySQL[7] and is made up of 3 tables:

- `City(id, Name, CountryCode, District, Population)`, which describe cities around the world;
- `Country(Code, Name, Continent, Region, SurfaceArea, IndepYear, Population, LifeExpectancy, GNP, GNPOld, LocalName, GovernmentForm, HeadofState, Capital, Code2)`, which describes countries;
- `CountryLanguage(CountryCode, Language, IsOfficial, Percentage)`, which lists the languages spoken in each country, together with the percentage of the population that speaks the language.

Join the tables `City` and `Country` to show a result where each city population is shown next to the country's population, for the country where the city is. Hint: `CountryCode` in `City` is a foreign key.

Exercise 3.6 Join the tables `CountryLanguage` and `Country` to show a result where each country name is shown next to the country's languages. Hint: here again attribute `CountryCode` is a foreign key.

The dot notation is necessary because sometimes name clashes are unavoidable. As we will see later, there are some occasions where we want to combine a table with *itself* : we want to take the CP or the join of a table T with T. What does this mean, and how is it accomplished? This simply means we take two *copies* of the data in T and take the CP or join of these two copies. But now *all* attributes are ambiguous! This problem is resolved by renaming the tables themselves, as the following example shows.

Example: Self-Join and Self-Product

The following query produces the CP of table T with itself:

```
SELECT T1.A, T1.B, T1.C, T2.A, T2.B, T2.C
FROM T AS T1, T AS T2;
```

[7] https://dev.mysql.com/doc/index-other.html.

resulting in table:

CP of **T** (as T1) and T (as T2)					
T1.A	**T1.B**	**T1.C**	**T2.A**	**T2.B**	**T2.C**
a_1	b_1	c_1	a_1	b_1	c_1
a_1	b_1	c_1	a_2	b_2	c_2
a_1	b_1	c_1	a_3	b_3	c_3
a_2	b_2	c_2	a_1	b_1	c_1
a_2	b_2	c_2	a_2	b_2	c_2
a_2	b_2	c_2	a_3	b_3	c_3
a_3	b_3	c_3	a_1	b_1	c_1
a_3	b_3	c_3	a_2	b_2	c_2
a_3	b_3	c_3	a_3	b_3	c_3

Note that the result has 9 rows (3×3) and that each row is combined with a copy of itself. The following query takes the join of table T with itself using a condition that requires equality on attribute A:

```
SELECT *
FROM T AS T1, T AS T2
WHERE T1.A = T2.A;
```

and results in table:

Join of **T** (as T1) and T (as T2)					
T1.A	**T1.B**	**T1.C**	**T2.A**	**T2.B**	**T2.C**
a_1	b_1	c_1	a_1	b_1	c_1
a_2	b_2	c_2	a_2	b_2	c_2
a_3	b_3	c_3	a_3	b_3	c_3

The result is a subset of the CP.[8]

As in the case of CP, it is possible to join 3 or more tables using a sequence of pair-wise join (in any order, since joins are also associative). In particular, a table can be joined with itself an arbitrary number of times by renaming the table as many times as needed. We will see a use for this later.

Exercise 3.7 Write a query that joins table T with itself 2 times.

[8] Since, as stated above, a join is a CP followed by a selection, the join is always a subset of the CP.

3.1.2 *Functions*

Functions are one of the most useful features of any database system. Unfortunately, the SQL standard specifies only a few, basic functions; most systems offer quite a few functions beyond these basic ones. The good side of this is that a rich functionality is available from most databases; the bad side is that the functions offered, even when they do the same thing, tend to vary in name and other details from system to system (and different systems may offer different functions). But most databases tend to cover the same basic operations in a similar way and therefore what is presented here is doable in pretty much any SQL database.

Functions in SQL can be used in the SELECT and in the WHERE clause. They are always applied to an attribute. There are two types of functions in SQL:

- (Standard) functions. These are functions that are applied to the value of some attribute in a row and yield a (single) result. Since most attributes are of type string, numerical, or date, the typical system has string functions (that is, functions that take a string value, and perhaps some additional parameters, and manipulate the value in certain ways), numerical functions (including typical arithmetic functions), and date functions (that is, functions that take a date value, and perhaps some additional parameters, and manipulate the date in certain ways). We introduce these functions in an as-needed basis throughout this chapter and the next one.

  We will use several mathematical functions in what follows. The most common ones are:

  - `POW(m,n)`, which calculates the exponentiation function m^n. It is important to note that in many systems there is a function `SQRT(m)` to take the square root $\sqrt{m}$, but for higher roots it is common to use `POW(m, 1.0/n)` (since $\sqrt[n]{m} = m^{\frac{1}{n}}$).
  - `LOG(m, b)`, which computes its inverse, the logarithm function (using base b);[9]
  - `MOD(m,n)`, which returns the remainder of dividing m by n (and is often used to compute modular arithmetic);
  - `CEIL(m)`, which returns the *ceiling* of m (that is, the least integer that is greater than or equal to m).

  These functions are available in Postgres, MySQL, and most other systems. There are also functions that deal with string and dates; a few are mentioned in the next example, and they are discussed in much more detail in Sects. 3.3.1.2 and 3.3.1.3.

---

[9] Confusingly, MySQL uses `LOG(base, number)` instead. Also, in Postgres a special notation exists for base e (`ln()`) and when `log` is used with a single argument, the default is base 10, while it is base e in MySQL.

- Aggregate functions. These are functions that take in a *collection* of values and produce a single result. *Aggregate functions can only be used in the SELECT clause*. The reason is that any function used in the WHERE clause as part of a condition applies to a single value (since it is evaluated in a single row), while aggregate functions, as stated, apply to collections. The SQL standard only requires the functions AVG (average or mean), SUM (sum of values), COUNT (count number of values), MIN (minimum), and MAX (maximum). However, most systems offer a large number of additional functions, including some statistical functions that we will put to a good use in the next sections (for instance, most systems provide *standard deviation*, usually called std or stddev).

Some examples will show the difference in meaning and use.

Example: Standard Functions

In the query

```
SELECT to_date(concat_ws("-", year, month, date))
           as flight-date,
       distance/1000.0 as distance-thousands,
       carrier || tailnum as plane-identifier,
       trim(both from flight) as flight-number,
       lower(dest) as destination
FROM ny-flight;
```

there is no WHERE clause, which means that all rows in table ny-flight qualify for the answer. On each row, the system will pick the value of attributes year, month, date, distance, carrier, tailnum, flight, and dest and on each row will use these values as follows:

- The function concat_ws takes several strings as arguments: the first one denotes a string that acts as a 'separator,' that is, it is used in between all other arguments as they are *concatenated* or composed into a single string. Thus, the function call above will put together the values of attributes year, month, and date in one string, but with a hyphen (the first argument) in between them.
- The function to_date takes the result of the previous function, which is a single string, and transforms it into a value that the system recognizes as a date. Note how this function takes as input the output of another function; *composition* of functions is fine as far as the output type of a function is the input type of another one. The result of this is called flight-date.
- The '/' is used to denote division, so this is an arithmetic function. The value of attribute distance is expressed in thousands after being divided by 1000. Note that the number is written as '1000.0'; the reason for this is that the system will interpret such a number (with a decimal point) as a real, not an integer, and '/' will act as real division (between two integers, '/' acts as integer division and provides always an integer result). A common trick when doing arithmetic is to divide or multiply an integer value by 1.0, thereby making sure that all other

operations (division, etc.) produce real numbers as result and no information is lost. The result of this division is called `distance-thousands`.

- The function'||' is another way of expressing concatenation in Postgres. Note that the values of attributes `carrier` and `tailnum` are put together as they are, without separators, so values "UA" and "N14228" end up as "UAN14228." The result is called `flight-identifier`.
- The function `trim` gets rid of whitespaces in a string. Recall that strings are enclosed in (double) quotes, and anything in between the quotes is part of the string. For instance, the string " UA " has 2 empty spaces before the non-whitespace characters "UA" and 2 more after it; they are all part of the string. In some cases, this causes confusion, as strings that intuitively seem the same are considered different by the system, due to the whitespaces. It is common to attack this problem by eliminating all whitespaces from a string; most system allow eliminating only whitespaces `before`, `after`, or `both` (as above) the non-white characters in the string. The result of the trimming is called `flight-number`.
- The function `lower` transforms all alphabetic characters in a string to their lowercase counterparts. Non-alphabetic characters and letters already in lowercase are left unchanged. Note that, to the computer, "Jones" and "JONES" are different strings. Hence, functions like `lower` are often used to make sure all values are expressed in the same way (this task, called *normalization*, is discussed in more detail in Sect. 3.4). The result is called `destination`.

The above is just a small sample showing some string, numeric, and date functions, using Postgres notation. In MySQL, concatenation of strings is also expressed using `CONCAT_WS` (if a separator is required) or `CONCAT` (if no separator is required) and conversion of strings to dates with `STR_TO_DATE`; trimming is written as `LTRIM` (for leading spaces) and `RTRIM` (for trailing spaces); converting all characters to lowercase is accomplished with `LOWER`. As we can see, the differences are minor.

Exercise 3.8 In the `world` database, calculate the number of years that a country has been independent (hint: there is an `IndepYear` attribute giving the year of independence, and all systems have functions that will tell you the current date and year).

Example: Aggregate Functions

Using table `Chicago-employees`, we want to calculate the average pay for salaried workers.

```
SELECT avg(salary)
FROM chicago-employees
WHERE salaried = 'T';
```

the system will first filter out rows according to the condition `Salaried = 'T'`; then, it will take the values of attribute `salary` in all the rows left and will apply aggregate function `avg` to them. The result is a single number.[10]

It is obvious now why aggregate functions cannot be used in the WHERE clause; it does not make sense to ask for the average, sum, or count of a single value.

Exercise 3.9 Count how many data points (flights) there are in the `ny-flights` table.

Example: Number of Unique Values

DISTINCT can be combined with aggregates like COUNT to find out the number of unique values in an attribute. The query

```
SELECT count(DISTINCT carrier)
FROM nyc-flights;
```

will tell us the number of airline companies mentioned in our New York flight database. Note the syntax: the DISTINCT keyword goes inside the parenthesis.

One of the characteristics of SQL that drives data analysts mad is that it takes a query to calculate an aggregated value. If you want to *use* it, you will have to write an *aggregated subquery*. Suppose, for instance, that you have the table *Chicago employees*, and you want to find out, among the salaried employees, which ones make more than average. One query is needed to compute the average salary, and another one to use this result and compute the answer.

Example: Aggregated Subquery

The query

```
SELECT name
FROM chicago-employees, (SELECT avg(salary) AS stat
                         FROM chicago-employees
                         WHERE salaried = 'T') AS Temp
WHERE salary > stat;
```

is evaluated as follows: the subquery in `FROM` is run, producing a result (which we can think of as a table named `Temp` with a single row and a single attribute). Following the idea of CP, this attribute is added to every row of table `table-name`,

[10]Technically, all SQL queries return a table as a result; therefore, for this kind of queries (sometimes called *aggregated queries*) where the system is guaranteed to return a single value, the result is technically considered as a table with a single attribute in the schema and a single row in the extension.

that is, it forms an additional attribute (with name `stat`). It is this 'extended' table that is evaluated in the `WHERE` clause; that is why the condition there refers to attribute `stat` as if it were an attribute of table `Chicago-employees`.

Subqueries that return a single result are very common since, as we stated above, SQL requires a query to compute an aggregate. Therefore, if the aggregate is to be used for further computation, a pattern like the one above is commonly used to express this.

Exercise 3.10 In the `world` database, give the name of countries with a GNP greater than the average GNP (for all countries).

Another annoying characteristic of aggregates is that calculating more than one aggregate at the same time can be sometimes difficult. For situations where we want to do several related calculations at once, SQL provides the CASE statement. The CASE is a *conditional* statement: its syntax is

`CASE WHEN` *`condition`* `THEN expression1 ELSE expression2 END`

and it means: evaluate the *condition*; if it is true, do `expression1`; if it is false, do `expression2`. It is often used in combination with aggregates, as the next example shows.

Example: Aggregates with CASE

Assume a table `Temperatures(day,month,temp)` that gives a temperature reading on a certain date, indicated by the month (1–12) and the day within the month. If we want the highest and average temperatures for January (1), this is easy to do:

```
SELECT max(temp), min(temp)
FROM Temperatures
WHERE month = 1;
```

Note that the WHERE filters the temperatures for January for both aggregates. However, if we want to have the highest temperatures for January (1) and also the highest for February (2), we have an issue. Which temperatures are used to compute a result is determined by the conditions of the WHERE clause of which there is only one. One solution is to write two queries, one for January and one for February. The first one is

```
SELECT max(temp)
FROM Temperatures
WHERE month = 1;
```

The second query is identical, except that the WHERE condition is `month = 2`. Experienced SQL programmers would write this as a single query, as follows:

```
SELECT max(CASE WHEN month = 1 THEN temp ELSE 0 END)
          as JanMax
       max(CASE WHEN month = 2 THEN temp ELSE 0 END)
```

```
        as FebMax
FROM Temperatures;
```

This can be read as follows: since there is no WHERE clause, all rows in table `Temperatures` are fed to the aggregates. However, on each row the system runs the CASE statement *before computing the aggregate*. In the first statement, we compare the value of the `month` attribute to 1; if it is, we pass the value of the attribute `temp` on that row to the `max` aggregate; otherwise was pass 0. This means that the max will only consider the temperatures of the rows where the month is 1. The same happens for the second `max` aggregate, but for `month = 2`.

This trick is used to do in one query what would otherwise require two queries. When the table extension is large, this can save not only some typing but also some time.

Exercise 3.11 For all flights originating at LGA, calculate the average duration ("air_time") when the flight is less than 500 miles and the average duration when the flight is longer than 500 miles.

There is a subtlety involving the computation of averages (means): we cannot write

```
avg(CASE WHEN month = 1 temp ELSE 0 END)
```

to compute the average since this would have taken the average of all January temperatures *and* a bunch of zeros. Instead, we should use

```
avg(CASE WHEN month = 1 temp ELSE NULL END)
```

since null markers are ignored by aggregate functions.

Example: Percentages

A very common use of the CASE pattern is to compute *percentages*. Assume that, using table `Chicago-employees`, we want to know what percentage of all money paid to hourly workers goes to employees in aldermanic duties. In SQL, we would compute (a) the sum of all money paid to hourly workers; (b) the sum of all money paid to hourly workers who are in aldermanic duties; the final result is the latter divided by the former. To do all in a single query, we can use

```
SELECT sum(CASE WHEN type = 'aldermanic-duties' wages ELSE 0)
       / (sum(wages) * 1.0)
FROM chicago-employees
WHERE hourly-wages = 'T';
```

Note that the condition in the WHERE clause affects both aggregates, since the WHERE clause is executed before the SELECT clause. Thus, both sums only take into account employees with wages, not salaried ones. However, among those, the CASE makes the first sum to only add the wages if the occupation type of the employee is aldermanic duty, while the second sum simply adds all the wages. The division then gives us the percentage we are after (we are using again the trick of multiplying by 1.0 to make sure that '/' is real division).

Example: Probabilities

Another very common use of this pattern is to approximate *probabilities*. Taking the data in a table as a sample of some underlying population, we can calculate probabilities for certain events by counting, out of all possible cases, which ones are of the form that we are interested in. Assume, for instance, that we want to know how many workers are employed to cover 'aldermanic duties'; we can write

```
SELECT sum(CASE WHEN type = 'aldermanic-duties' THEN 1
                ELSE 0)/(count(*) * 1.0)
FROM chicago-employees;
```

In this example, the first sum adds one for each row where the type is 'aldermanic-duties' and ignores others (effectively counting how many times the type is 'aldermanic-duties'), while the count counts all the rows. The division then provides a percentage or estimated probability.

The pattern from these examples can be generalized: the probability that event A happens, among all the population, is given by

```
SELECT sum(CASE WHEN A THEN 1 ELSE 0)/(count(*) * 1.0)
FROM table;
```

The conditional probability that event A happens, given that event B has happened, is given by

```
SELECT sum(CASE WHEN A THEN 1 ELSE 0)/(count(*) * 1.0)
FROM table
WHERE B;
```

Exercise 3.12 In the ny-flights table, all flights originate in one of 3 airports: 'JFK' (Kennedy), 'LGA' (La Guardia), and 'EWR' (Newark in New Jersey). Count how many flights originate at 'JFK.' Then show how many flights originate at 'JFK' as a percentage of all flights.

3.1.3 Grouping

We have seen how to calculate aggregates over the whole table or subsets of the whole table. However, it is very common in data analysis to be interested in some statistic (be it raw counts, or means, or something else) for different categories within the same dataset (especially using categorical attributes). Because this is a very common pattern across scientific and business fields,[11] SQL has a set of tools specifically designed for this purpose. The GROUP BY clause tells the system to partition the table into groups: then, any aggregates or actions in the SELECT clause are carried out within each group. The table returned as final result contains *one row per group*. This row contains whatever aggregates or information we have requested in our SELECT for each group.

Example: Split-Apply-Combine in SQL

The following gives the number of flights out of JFK for each airline.

```
SELECT carrier, count(*)
FROM ny-flights
WHERE origin = 'JFK'
GROUP BY carrier;
```

Here, the system scans the table `ny-flights` and picks the rows where the `origin` attribute has value 'JFK'; this set of rows is fed to the GROUP BY. In this case, `GROUP BY carrier` will break the set into groups of rows; each group will be composed of all the rows that share the same value for attribute `carrier`—note that the number of groups, and the number of rows on each group, depends on the particular dataset. Once this partition is done, the system will apply a `count(*)` aggregate *to each group*. Finally, the system will return as answer a table with two attributes, the first one the carrier name, and the second one the number of rows in the group representing that carrier name. In other words, it will tell us how many flights there are in the dataset for each airline, but counting only the flights that originate at JFK. Importantly, for this to work we assume that each flight is represented by one row in the table. Under different conditions, the query would have to be written differently.

Note that when the GROUP BY is combined with a WHERE clause, the WHERE is always done first and it filters the rows that are available to the GROUP BY to do its partitioning.

Exercise 3.13 As stated earlier, all flights in the `ny-flights` table originate in one of 3 airports: 'JFK' (Kennedy), 'LGA' (La Guardia), and 'EWR' (Newark in New Jersey). Count how many flights originate at each one of these airports.

Exercise 3.14 Using the `world` database, count the number of languages spoken in each country. Identify the country by its code (no join needed). Repeat identifying each country by name (join needed).

One of the most typical uses of this clause is the creation of *histograms*: a histogram gives us the frequency associated with a value, i.e. the number of times a value appears in a dataset.

[11] It is sometimes given the name *split-apply-combine* [18].

Example: Histograms

Suppose we want to know the associated frequency to each destination airport. Then we would write

```
SELECT dest, count(*) as frequency
FROM ny-flights
GROUP BY dest;
```

This would return a table with all destination airports and, for each one, the number of times (raw frequency) from which it appears in the dataset (under the assumption that one flight = one row, the frequency represents the number of flights into that airport).

The GROUP BY clause takes a list of attribute names; in fact, expressions that return an attribute name, like a CASE, can be used. When more than one attribute is mentioned, then all the values involved are used to determine the groups that data is broken into, as the next example shows.

Example: Complex Grouping

Suppose we are interested in the number of different ways to get from one airport to another. The query

```
SELECT origin, dest, count(*) as joint-frequency
FROM ny-flights
GROUP BY origin, dest;
```

breaks the table into groups using attributes `origin` and `dest`: that is, to be in the same group, two rows must have the same value in both attributes. Hence, each group represents all the flights that take us from `origin` to `dest`.

As before, we can convert raw counts to percentages, but we need to be careful. The frequency should be divided by the total number of data points, i.e. the number of rows in the whole table. However, aggregates computed withing grouping apply to each group. Therefore, we need to compute this total number apart (in a subquery). When we attempt to use the total, we may run into trouble with SQL. The reason is that using a GROUP BY restricts what can be extracted out of the data in the SELECT clause. In any SQL query with a GROUP BY, the SELECT can only mention attributes that appear in the GROUP BY or aggregations. This is because the result of a query is a table consisting of one row per group, so only expressions that guarantee single values per group are allowed. This restriction can get on the way of analysis sometimes; as we will see in Sect. 5.3, SQL provides a more flexible way of partitioning data that lifts this restriction.

Example: Incorrect Grouping

In table ny-flights, I want to see the number of flights per destination airport and also the airlines that provide those flights, so I write

```
SELECT dest, carrier, count(*)
FROM ny-flights
GROUP BY dest;
```

This query is **not** legal SQL and will generate an error.[12] The reason is that when the system attempts to generate one tuple per group, it finds out that there is only one value of `dest` (since this is the attribute used to create the group, all tuples in the group have the same value for it) and only one value of `count(*)` (since it is an aggregate) but there are multiple values of `carrier`, and room for only one. One way to get this information is to split the query into two and get the information separately; if necessary, both answers can be combined into one:

```
SELECT T1.dest, T1.carrier, T2.num
FROM ny-flights as T1, (SELECT dest, count(*) as num
                        FROM ny-flights
                        GROUP BY dest) as T2
WHERE T1.dest = T2.dest;
```

This table will show one row for each carrier–destination combination and repeat the number of flights for that destination in all tuples with the same destination.

This limitation of SQL syntax can sometimes be aggravating, as we see next.

Example: Grouping and Percentages

If we wanted to convert the table where we counted the numbers of flights per destination into percentages, we need to divide by the total number of flights. It would seem that the way to do this is as follows:

```
SELECT dest, count(*) / (total * 1.0)  as probability
FROM ny-flights,
     (SELECT count(*) as total FROM ny-flights) AS T
GROUP BY dest;
```

This query uses a *subquery* in the FROM clause to count the number of rows in the whole table. One way to see what is happening is to imagine that this subquery is evaluated before anything else. It produces, as a result, a table with a single row and a single column. The system then takes the Cartesian product of this table and table `ny-flights`. Since we have only one row in the table, the end result is to attach, to each row in `ny-flights` a new value named `total`, which contains the result of

[12]Early versions of MySQL would return a non-sensical answer instead of giving an error.

the subquery (note that this single value is repeated in each row!). However, most systems will actually throw an error with this query. This is because the attribute `total` will be seen as a non-aggregated, non-grouping attribute of the table being grouped (which is the CP of `ny-flights` and `T`). There are two work-arounds for this. The first is to observe that the value of `total` is actually the same in all rows; hence, we can write

```
SELECT dest, count(*) / (min(total) * 1.0)  as probability
FROM ny-flights,
     (SELECT count(*) as total FROM ny-flights) AS T
GROUP BY dest;
```

By applying the aggregate `min`, we have converted attribute `total` into an aggregated result. Since the value of `total` is the same in all rows, applying `min` to it does not change its value. The second approach is to compute the value as follows:

```
SELECT dest, sum(1.0 / total) as probability
FROM ny-flights,
     (SELECT count(*) as total FROM ny-flights) AS T
GROUP BY dest;
```

Here we are adding the individual 'weight' of each fact, with the 'weight' indicating how much a single fact counts ($\frac{1}{n}$ in a dataset with n facts). Note that, in all cases, we use the trick of multiplying by 1.0 to make sure that floating number division, not integer division, is used.

Exercise 3.15 Turn the joint histogram above into a joint probability distribution of attributes 'origin' and 'destination' by converting the raw counts in the example above to percentages/probabilities, using the same approach as the previous example.

We can combine GROUP BY with the use of CASE in the definition of the groups, so that we can make up groups more generally and not just by value.

Example: GROUP BY and CASE

The following query counts how many flights are small, medium, and large distance, according to some prefixed cut points:

```
SELECT  CASE WHEN distance < 500 THEN 'Short'
             WHEN distance < 100 THEN 'Medium'
              ELSE 'Long' END as flightDuration,
              count(*)
FROM ny-flights
GROUP BY flightDuration;
```

Of course, CASE can also be used in the aggregates of a group-apply-combine query.

Example: GROUP BY, Aggregates, and CASE

The following query will count the number of short, medium, and long flights per destination, according to the same cut points as before:

```
SELECT dest
    sum(CASE WHEN distance < 500 THEN 1 ELSE 0 END)
         as shortFlights,
    sum(CASE WHEN distance BETWEEN 500 and 1000
              THEN 1 ELSE 0 END) as  mediumFlights,
    sum(CASE WHEN distance > 1000 THEN 1 ELSE 0 END)
         as LargeFlights
FROM ny-flights
GROUP BY dest;
```

Exercise 3.16 In `ny-flights`, count the number of flights with a departure delay of less than 5 min, a departure delay between 5 and 10 min, and a departure delay of more than 10 min.

An additional tool for this type of analysis is the HAVING clause. This clause can only be used with GROUP BY, never on its own. The reason is that HAVING puts a condition on partitions; that is, it examines partitions created by GROUP BY and it lets them through or filters them out.

Example: HAVING Clause

Suppose that, as before, we want to know the associated frequency to each destination airport but are only interested when the frequency is higher than some threshold, say 10. Then we would write

```
SELECT dest, count(*)
FROM ny-flights
GROUP BY dest
HAVING count(*) > 10;
```

This query would be executed as the previous example, with the table being partitioned by destination, and the aggregate applied to each partition. However, the system will only return a row for those partitions where the aggregate `count(*)` generates a value greater than 10.

Queries with grouping are very useful for examining overall tendencies of the data, especially distribution of values; HAVING clauses are useful in narrowing down such an examination.

Exercise 3.17 In `ny-flights`, count the number of flights with an arrival delay of more than 15 min for each airline but show only airlines where that number is at least 5 flights.

Exercise 3.18 In the `world` database, show the number of cities with more than 100,000 inhabitants for each country but show only countries with at least 5 such cities.

Exercise 3.19 In the `world` database, show the number of languages per country but show only countries where more than 2 languages are spoken.

Exercise 3.20 In the `world` database, show (for each country) the number of languages spoken by at least 10% of the population of that country but show only countries with more than 2 such languages.

3.1.4 Order

So far, all the queries we have seen return tables as answers. One of the characteristics of the table is that row order is not important, that is, an answer is a *set* of rows: the only thing that matters is which rows made to the answer. However, sometimes we want more than that. In particular, sometimes we want results that are *ordered*. For these cases, SQL has an additional clause, `ORDER BY`, which allows us to impose an order on the result of a query. The `ORDER BY`, if used, is added after any `WHERE` or `GROUP BY` clauses.

Example: ORDER BY Clause

Assume (once again) that we want to know, in the `ny-flights` dataset, about the number of flights per destination. However, while we plan to examine all destinations, we would like to be able to see the most frequent destinations first. We can write

```
SELECT dest, count(*)
FROM ny-flights
GROUP BY dest
ORDER BY count(*) desc;
```

The keyword `desc` stands for *descending order*; `asc` can be used for ascending order (but it is the default, so that is usually not necessary).

Exercise 3.21 In the `world` database, sort the countries by number of languages spoken (most languages first).

Ordering can be used in conjunction with another clause, `LIMIT`. This clause takes a number as its argument: `LIMIT n` means that the system will return at most n rows as answer. Note that using `LIMIT n` by itself will simply truncate an answer; most of the time, `LIMIT` is combined with `ORDER BY` to answer *top k* queries: these are questions when we want the most important *k* elements in the answer, not all of them, with 'importance' decided by some aggregate or measure defined in the query.

Example: Top *k* Query

The query

```
SELECT dest, count(*)
FROM ny-flights
GROUP BY dest
ORDER BY count(*) desc;
LIMIT 10;
```

will retrieve the top 10 destinations by number of flights. The query

```
SELECT dest, count(*)
FROM ny-flights
GROUP BY dest
ORDER BY count(*) desc;
LIMIT 10 OFFSET 20;
```

will pick destinations 21 to 30 (10 destinations, starting after 20) in the order created by the count.

Top *k* queries are especially useful for distributions where we can expect a *long tail*, that is, a long list of marginally useful results. This is typical in scenarios where can expect a *power law* distribution of results: for instance, an analysis of computers in a network may show a few computers that handle most of the traffic (the servers), while a bunch of computers have only limited traffic (individual PCs). Another typical example is wealth distribution: there are, in most countries, a few billionaires, a small number of millionaires, followed by a very large number of people with limited income. When we want to focus on the 'top group' and avoid the long tail, a top *k* query is our friend. Note that it is trivial to get "bottom *k*" results by using ascending instead of descending order.

However, `LIMIT` can be used by itself, in which case it can be seen as a rudimentary form of *sampling*), since the rows to be retrieved are picked by the system. `LIMIT` can also be combined with `OFFSET` to pick not just the top *k*, but any *k* answers: `OFFSET n` makes LIMIT to start counting at row $n + 1$ instead of at row 1.

Example: Top *k* Query

The query

```
SELECT dest, count(*)
FROM ny-flights
GROUP BY dest
ORDER BY count(*) desc;
LIMIT 10 OFFSET 20;
```

will retrieve destinations 21 to 30, with the order coming (as in the previous example) by number of flights.

Exercise 3.22 In the `world` database, show the 10 most populous cities.

Exercise 3.23 In the `ny-flights`, show the flight origin, destination, and airline with the top 10 flights by delay (i.e. largest arrival delay).

Ordering can be also made explicit with *ranking*, whereby each row gets an additional attribute that gives its order within the result. We discuss how to produce and use rankings in Sect. 5.3.[13]

3.1.5 Complex Queries

We have seen above several examples where a subquery in the FROM clause is used to help obtain a result. Using such subqueries is not uncommon; sometimes, the SQL syntax forces us to break down a computation into steps, each step requiring a query. A typical example, as already observed, is the use of aggregates, which need to be computed first.

Using subqueries can be helpful when writing complex queries, as they allow us to break down a problem into sub-problems. Because of this, SQL has several ways in which subqueries can be used. Besides subqueries in FROM, another useful construct is the WITH clause. This clause precedes a query and introduces a temporary table that can then be used in the query. Its syntax is

```
WITH table-name AS (SELECT ... FROM ... WHERE ...)
SELECT ...
FROM table-name, ...
WHERE ...
```

The first query (in parenthesis) defines a table that is given a name (and optionally, a schema); this name can then be used in following query. This is equivalent to defining a subquery in the FROM clause:

```
SELECT ...
FROM (SELECT ... FROM ... WHERE ...) as table-name, ...
WHERE ...
```

but is preferred by some programmers as it makes more clear that we are creating a temporary table for further computation.

[13]Rankings can be computed without the advanced methods of Sect. 5.3, but doing so is quite costly and it should not be done except with small datasets.

Example: Subqueries in FROM, Revisited

In a previous example, we ask where those flights coming into JFK on December 10 are coming from, but only for trips longer than 1,000 miles:

```
SELECT origin
FROM (SELECT id, distance, origin
      FROM ny-flights
      WHERE year = 2013 and month = 11 and
            day = 10 and dest = "JFK") AS T
WHERE distance > 1000;
```

This could also have been written as

```
WITH T AS
 (SELECT id, distance, origin
  FROM ny-flights
  WHERE year = 2013 and month = 11 and
        day = 10 and dest = "JFK")
SELECT origin
FROM T
WHERE distance > 1000;
```

Another reason to use WITH is that it can be combined with subqueries in FROM clause for complex cases where several queries are required.

Example: Complex Queries

Suppose a company has a table `ASSIGNMENTS(pname, essn, hours)`, which states which employees (denoted by `essn`) are working in which projects (denoted by `pname`) for how many `hours` a week. The boss wants to know which project(s) require the largest number of person-hours. In order to compute this in SQL, we need to take the following steps:

1. Compute the number of person-hours per project.
2. Find out the largest number of person-hours.
3. Find the project(s) with that largest number.

This requires 3 queries, which we can put together as follows:

```
WITH PERSON-HOURS(pname,total) AS
 (SELECT pname, sum(hours)
  FROM ASSIGNMENTS
  GROUP BY pname)
SELECT pname
FROM PERSON-HOURS,
     (SELECT max(total) as maxt FROM PERSON-HOURS) AS T
WHERE total = maxt;
```

The WITH clause is evaluated first, creating a temporary table `PERSON-HOURS`, which is then used in evaluating the subquery in the FROM clause of the second query and the rest of the second query.

Breaking down an answer into steps, writing a query for each step, and then combining the queries to obtain a final result is a good tactic when using SQL. In the following, we will use WITH and subqueries in FROM liberally to solve problems.

Exercise 3.24 Let the *linguistic diversity* of a country be defined as the number of languages spoken in the country divided by the country's population expressed in millions. Find the country(ies) with the largest linguistic diversity in the database.

3.2 Exploratory Data Analysis (EDA)

The first thing to do with a dataset is to find out its meaning and main characteristics. The meaning is whatever events or entities the tables refer to, and whatever attributes or characteristics the attributes denote—in other words, what information the data represents. The goals of Exploratory Data Analysis (EDA)[14] on a dataset are: to determine what type of dataset it is (structured, semistructured, or unstructured); if it is not unstructured, what is schema is—that is, what attributes compose the data (and, in the case of semistructured, how they are combined—what the tree structure is) and to determine, for each attribute, the type of domain it represents (categorical/nominal, ordinal, or numerical) and what the data values in the domain are like: what a typical value is, what the range of values is, and so on.

In the rest of this chapter, we will assume for now that we are dealing with structured data and that the schema is known. Hence, the main task will be to find out about each domain. In such an scenario, EDA will proceed as follows:

1. Examine each attribute in isolation. This is called *univariate* analysis in the statistical literature, the name indicating we are dealing with *one* variable at a time. The first task is to determine the type of domain. Typically, categorical attributes are represented by string data types, numerical attributes by number data types, and temporal information by times and/or dates (ordinal domains can be represented by strings or numbers, depending on the context). However, this is not always the case, and EDA should be used to find any mismatch between the domain type and the data type and correct it. Once we know the type of domain, we explore the values on it as follows:
   - For categorical attributes, histograms and its variations are the main tool.

[14]This step is sometimes called *data profiling* or *data exploration*.

- For numerical values, we want to find the measures of central tendency (mean, median, mode, and a few more) and dispersion (standard deviation, quartiles). We may also try to fit a known distribution to a domain to see if the domain can be described succinctly by a formula.

At this state, we should also try to identify potential issues on the domain of each attribute. Some of the main problems to look for are *missing values*, *outliers*, and errors (data in the wrong type, etc.). Discovering whether such problems are present in our dataset will help with the data cleaning step (see Sect. 3.3).

2. Examine the relationships (or lack thereof) among several attributes. This is called *multivariate* analysis in the statistical literature. It is common that some attributes in a dataset are related to other attributes; since each attribute describes an aspect or characteristics of an object or event, there is the possibility that two or more attributes exhibit some common traits or are connected in some way. The techniques for examining potential connections depend on the types of attributes involved; some basic tools that we will see later are (classified by variable type):

- For **categorical–categorical** analysis (both variables involved are categorical): contingency tables, chi-square test.
- For **categorical–numerical** analysis (one variable is categorical, the other one is numerical): logistic regression, ANOVA.
- For **numerical–numerical** analysis (both variables are numerical): covariance, correlation, PMI, linear regression.
- For **ordinal–ordinal** analysis (both variables are ordinal): Spearman's rank, Kendall's rank.

In principle, any set of schema attributes could be related to another set of schema attributes. Unfortunately, a schema with n attributes has 2^n sets of attributes, so the number of potential connections between these would be of the order of 2^{2^n}, a number that is too large even for small values of n. Therefore, EDA typically focuses on *pairs* of attributes A and B, trying to decide if there is any connection between the two of them. More complex connections can be also explored, but this usually happens during the data analysis phase itself (see next chapter).

The main tools of EDA are *visualization* and *descriptive statistics*. Visualization is very useful because people are very good at discerning patterns when these are presented in a graphical manner. For a single attribute, a *bar chart* of a *histogram* can give a quick overview of value distributions. *Boxplots* can also be used for exploration. The boxplot of a symmetric distribution looks very different from the one of an asymmetric distribution. For two attributes, a *scatterplot* or a *density plot* is useful. Unfortunately, visualization is a weak point of databases. Most database systems offer nothing and depend on third-party, external tools for visualization. Tools like R are much better at this (see Sect. 6.1 for an overview of R). We will not discuss visualization any further in this book. Fortunately, simple descriptive statistics can be computed in databases; in the rest of this section, we describe how to carry out EDA in SQL.

To summarize, EDA can be described as a set of tools that help us build a description of the dataset. The goal of EDA is to gain an intuition about what the data in the dataset is like, to summarize it and to try to identify any issues with it. This is purely a descriptive task, so assumptions made are minimal at this stage. EDA is very important because the information that we obtain at this stage will guide further work in the next stages, from data cleaning (Sect. 3.3) and data pre-processing (Sect. 3.4) to data analysis (Chap. 4).

3.2.1 Univariate Analysis

As stated, the first task is to determine whether the domain of the attribute is categorical, numerical, or ordinal. It should be the case that categorical attributes are expressed with some kind of string data type, and numerical attributes with some kind of number data type, but this does not need to be the case. For instance, products in a catalog can be divided into 5 categories, which are called (regrettably) '1,' '2,' '3,' '4,' and '5.' This is not a numerical attribute, not even an ordinal one (unless there is an underlying reason to use those numbers, for instance reflecting increased price range). It is also very common to have temporal information like dates entered as strings, instead of as a date data type. This should be avoided, since dates have their own functions that can be fruitfully used during analysis. Establishing the (intended) meaning of each attribute is essential here; this is what allows us to compare what we find to what could be expected.

We first consider numerical attributes. In this case, we are interested in the measures of the range (minimum, maximum), central tendency (different means —arithmetic, geometric and harmonic, median, mode), and dispersion (standard deviation, variance, skewness, kurtosis) [4, 5]. It is easy to calculate several of these values at once; for a given attribute `Attr` in table `Data`, we can use the following query to extract some basic information:[15]

```
SELECT count(Attr) as number-values,
       count(distinct Attr) as cardinality,
       min(Attr) as minimum,  max(Attr) as maximum,
       max(Attr) - min(Attr) as range,
       avg(Attr) as mean,
       stddev(Attr) as standard-deviation
FROM Data;
```

We examine each of these in more detail next.[16] However, it is important to point out right away that aggregate functions *skip nulls*, that is, if there are null markers in

[15]For readers familiar with R, note the similarity with the `summary` command.

[16]From this point on, we show SQL *templates*, that is, we use generic attributes and tables to show how queries should be written. For instance, in the example above `Attr` should be substituted by a particular attribute name, and `Data` by a particular table name.

the attribute `Attr` being analyzed, those null markers will be ignored (except when using `count(*)`, which simply counts the number of tuples, regardless of what the tuples contain). We deal with null markers in Sect. 3.3.2.

The first aggregate simply counts how many data points (i.e. number of rows) we have; the second counts how many `unique` values are present in the attribute. The difference between the minimum and the maximum value gives us the *range*. The (arithmetic) *mean* of an attribute (also called *average* in popular parlance, and *expectation* or *expected value* in statistics) in a dataset is simply the sum of the values divided by the number of values. In all SQL systems, as we have seen, there is an aggregate function, `avg()`, that calculates this:

```
SELECT Avg(Attr) as mean
FROM Data;
```

Note that, by definition, this is the same as

```
SELECT Sum(Attr) / (Count(Attr) * 1.0) as mean
FROM Data;
```

We are using once again the trick of multiplying by 1.0 to make sure the system uses floating point division, not integer division.

For readers who see this concept for the first time, it will become important later to note the connection with frequencies and probabilities: when we add all the values in an attribute, if value a appears n times, we will add it n times; hence, we could achieve the same by adding each distinct value once after multiplying it by its frequency:

```
WITH Histogram(Value, Frequency) AS
(SELECT Attr, count(*)
 FROM Data
 GROUP BY Attr)
SELECT (1.0 * Sum(Value * Frequency)) / Sum(Frequency) as mean
FROM Histogram;
```

And, since we end up dividing by the total number of values, we could substitute the frequency by its normalized value, the probability:

```
WITH NHistogram(Value, Prob) AS
 (SELECT Attr, sum(1.0/total)
  FROM Data,
       (Select count(*) AS total FROM Data) AS Temp
  GROUP BY Attr)
SELECT Sum(Value * Prob) as mean
FROM NHistogram;
```

The mean is known to be affected by *outliers*, extreme values that may or may not be correct. Assume, for instance, that in our demographic dataset we have a `height` attribute, giving the height of each person in the dataset in feet and inches. What happens if we find a very low (say, 4 feet 10 inches) or very high (say, 7 feet and 10 inches) value? They could simply reflect that we have a very short or very tall person, or they could be the result of a mistake in measurement. What is clear is that this value will have a strong influence on the value of the mean. For this reason, some people prefer to use the *trimmed mean*, which is calculated after disregarding extreme value (usually, the maximum and minimum). This can be generalized, if one wants, to *k% trimmed mean*, where the highest/lowest k% of the values is removed [8]. A simple trimmed mean is quite easy in SQL:

```
SELECT avg(A)
FROM Data, (SELECT max(A) as Amax FROM Data) AS T1,
           (SELECT min(A) as Amin FROM Data) AS T2
WHERE A < Amax and A > Amin;
```

Note that, since the `WHERE` part is run first, the maximum and minimum values are eliminated from the `avg(A)` computation. However, a k% trimmed mean is much trickier; we will see how to compute this in Sect. 5.3.

Sometimes we require other kinds of mean. The *geometric mean* is the *product* of the n values divided by the n-th root of the values. This mean has the advantage of not being as sensitive to outliers as the arithmetic mean (in particular, large values do not disturb it by much, although small values have a significant effect). SQL does not have an aggregate function for multiplication; the standard work-around for this is to use logarithms, as follows: since $\log(ab) = \log a + \log b$, we can calculate $ab = \exp(loga + \log b)$; this generalizes to more than two values. Thus,

```
SELECT exp(sum(log(Attr))
FROM Data;
```

will give us the product of the values of attribute `Attr`. For the geometric mean, we need to compute the n-th root, which we can achieve with

```
SELECT pow(exp(sum(log(Attr))), 1.0 /total)
FROM Data, (SELECT count(Attr) as total FROM Data) AS T;
```

However, the same can be achieved by applying the exponentiation to the arithmetic mean of the sum of the logs. In SQL,

```
SELECT exp(sum(log(Attr)) / count(Attr))
FROM Data;
```

Since the sum divided by the count is the average, this can be simplified to

```
SELECT exp(avg(log(Attr))
FROM Data;
```

However, one needs to be careful for this due to issues of *numerical stability*: real numbers (such the result of calculating logarithms) are represented in the database (in the computer, really) with a certain degree of precision; numbers that are very small (or very large) may get rounded. The logarithm of a small number will tend to be a very small number,[17] and as a consequence the result obtained from the query above may, in some cases, be inexact.

The geometric mean is useful when dealing with growth (or decay) rates. The typical example is calculating interest rates: suppose a bank proposes, for a savings account, to give different (increasing) interest rates depending on how long the money is left in the bank. For the first year, it will give a 1% interest rate, for the second, 1.5%, for the third, 2%, for the fourth, 2.5%, and for the fifth, 3%. The (geometric) mean of the interests is simply

$$gm = (1.01 \times 1.015 \times 1.02 \times 1.025 \times 1.03)^{\frac{1}{5}} = 1.0199.$$

Note that to calculate how much an amount, say $1000, grows, we normally would calculate

$$(1000 \times s1.01.015 \times 1.02 \times 1.025 \times 1.03) = 1103.94,$$

which is the same as $1000 \times gm^5$.

Another mean is the *harmonic mean*, which uses the sum of the reciprocals:

```
SELECT Count(Attr) / Sum(1/Attr)
FROM Data;
```

This mean is used for measures based on ratios (that is, any measure that depends on some unit) or for when different amounts contribute with a different weight to the mean. The typical example used to illustrate this context is calculating speeds: assume a car travels at 60 miles/hour a certain distance and then comes back, traveling at 30 miles/hour. The average speed is the harmonic mean, 40 miles/hour. The reason is that the car traveled the same distance on both speeds, but since the return speed was a third of the original one, the car took 3 times as long going back. If the car had traveled the same time at both speeds, its average would be the traditional (arithmetic) mean, 45 miles/hour.

Exercise 3.25 Assume a table called `Trips` with an attribute `speed` and write a query to compute the arithmetic, geometric, and harmonic mean of this attribute.[18]

[17] This depends on the base, of course; the above uses e as the standard base.

[18] Any exercise that starts with *Assume some data* ... will describe a (highly simplified) scenario. These exercises can be carried out in either MySQL or Postgres (or any other relational system) as follows: (1) create the table described in the scenario (in this scenario, a table called `Trip` with a numeric attribute called `speed`); (2) insert some made-up data in this table (usually, 4–6 rows are enough); (3) write the query for the exercise. This will make it possible to ensure that: (a) the query is syntactically correct (otherwise the system will give an error message instead of an answer); and (b) the query is semantically correct (the answer can be checked by hand against the data).

The *mode* is the most frequently occurring value. In SQL, it must be calculated in steps:

```
WITH Histogram as
(SELECT Value as val, count(*) as freq
FROM Data
GROUP BY Attr)
SELECT val
FROM Histogram, (SELECT max(freq) as top FROM Histogram) AS T
WHERE freq = top;
```

Note that there can be several modes in a dataset. When we know that there is only one mode, the following is more efficient:

```
SELECT Value, count(*) as freq
FROM Data
GROUP BY Value
ORDER BY freq DESC
LIMIT 1;
```

However, this is correct only when we know for sure the number of modes—if there are *k* modes, we can use `LIMIT k`. Otherwise, this approach is incorrect.

As we will see, in some systems it is possible to call an aggregate function to compute the mode directly using windows (see Sect. 5.3).

Exercise 3.26 In `ny-flights`, compute the most visited destination (the destination with more flights arriving to it).

Finally, the *median* is the value that would appear in the middle position if all values were ordered from smaller to larger. That is, the median has as many values larger than it as values smaller than it. While the concept is very simple, it is actually a bit difficult to calculate in SQL. First, it requires values to be sorted; second, it requires that the middle position be found. Note that, for an odd number of values, the middle position is clearly defined (for instance, for 21 values, position 11 has 10 values before and 10 values after it), but for an even number of values, there is no real 'middle': this is usually solved by averaging the 'before' and 'after' values (that is, for 20 values, we would take the average of the values in positions 10 and 11). As a consequence of this, finding the median in SQL is a bit elaborate; the most efficient way is to sort the values with ORDER BY and use a combination of LIMIT and OFFSET to find what we want:

```
SELECT avg(Attr)
FROM (SELECT Attr
      FROM Data, (SELECT count(*) as size FROM Data)
      ORDER BY value
      LIMIT 2 - MOD(size, 2)
      OFFSET CEIL(size / 2.0)) AS T;
```

The idea here is that when `size` is odd, `MOD(size, 2)` is 1 and `CEIL(size / 2.0)` will be the 'middle' so we will pick a single value, the one in the middle position; but when `size` is even, `MOD(size, 2)` is 0 and `CEIL(size / 2.0)` will be the 'middle' minus 1, so we will pick two values, corresponding to the 'before' and 'after' the middle position. Note that this approach does not ignore NULLs, as most aggregates do; to discard them, one should add a WHERE clause using the `is not null` predicate. A simpler way to calculate the median will be shown in Sect. 5.3.

As for the measures of dispersion, the simplest is the *range*, or difference between the maximum (largest) and minimum (smallest) value. We have seen that this is trivial to calculate in SQL. The *standard deviation* is probably the most common dispersion measure. It is also easy to calculate, since it is a built-in function in most database systems:

```
SELECT stddev(Attr) FROM Data;
```

is how Postgres expresses it. In MySQL, the function is call `std`.[19]

It is instructive, though, to look at how to calculate this value. The typical formula for standard deviation for some collection of values x_i is

$$\sqrt{\frac{\Sigma_{i=1}^{n}(x_i - \mu)^2}{(n-1)}}$$

with μ the mean. In SQL, this yields

```
SELECT sqrt((sum(power(Attr - mean, 2)) / (count(*) -1))
FROM Data, (SELECT avg(Attr) as mean FROM Data) as T;
```

The following formula is equivalent to it and is easier to compute (it requires only one pass over the data):

$$\sqrt{\frac{(\Sigma_{i=1}^{n} x_i^2)}{(n-1)} - \left(\frac{(\Sigma_{i=1}^{n} x_i)}{(n-1)}\right)^2}$$

In SQL:

```
SELECT sqrt( sum(pow(Attr, 2)) / (count(*) - 1) -
             pow(sum(Attr) / (count(*)-1), 2)) as std-dev
FROM Data;
```

The value of this example is to show that, in many cases, when a function is not available in the system, it is still possible to produce a result by implementing its definition—something we exploit often in this chapter and the next.

[19]The function `stddev()` in Postgres is an alias for `stddev_samp` (standard deviation over the sample), which divides (as we do here) over $n - 1$ when there are n data values. There is also a `stddev_pop` function that divides over n. Similar functions exist in MySQL.

In some cases we will need the *variance*, which is simply the square of the deviation. Variance is also available as a built-in in most systems; in both Postgres and MySQL, one can use:[20]

```
SELECT variance(Attr) FROM Data;
```

Exercise 3.27 Pretend the variance is not available in your system; compute it from scratch using the approach for standard deviation.

Both standard deviation and variance are also affected by outliers, so sometimes the MAD (Median Absolute Deviation) is calculated too. The MAD of a set of values is the median of the absolute value of the difference between each value and the median:

$$MAD(x_i) = \text{median}(|x_i \ - \ \text{median}_i(x_i)|)$$

It can be seen as a deviation measure that uses the median instead of the mean, and the absolute value instead of the square root. Using the median as calculated above, this formula can be transformed into an SQL query.

Exercise 3.28 Give the SQL to calculate the MAD of attribute `Value` in table `Data`. Hint: reuse the definition of median above; use a WITH clause.

Exercise 3.29 When data can be grouped, it is common to examine the *between-classes variance*: if attribute `A` in table `T` can be divided into groups $g_1, \ldots, g_n$, let n_i be the size of interval g_i and avg_i the mean of the values in interval g_i. Also, let avg be the mean of all `A` values and m the total number of `A` values. Then

$$\frac{1}{m}\Sigma_{i=1}^{n} n_i(avg_i - avg)^2$$

is the between-class variance. Transform this formula into an SQL query. Hint: use subqueries in `WITH` and/or `FROM` to compute the intermediate results needed (the means within each interval, the number, and mean of all values).

Exercise 3.30 If we add all the between-class variances for a variable, we get less than the total variance, because we disregard the variance *within* the class. However, with access to the raw data we can actually make up for this and compute the variance as the sum of the between-class variance and the *within-class* variance:

$$\frac{1}{m}\Sigma_{i=1}^{n} n_i(avg_i - avg)^2 + \frac{1}{m}\Sigma_{i=1}^{n} n_i var_i^2$$

[20] As in the case of the standard deviation, there are functions `var_samp` and `var_pop` in both systems.

where var_i is the variance in interval g_i. Transform this formula into an SQL query. Hint: the first part is the same as the previous exercise; reuse all your work in a subquery. The second part can also be calculated with a separate subquery.

Sometimes a couple of additional statistics can prove useful: the *third moment* about the mean or *skewness*, and the fourth moment about the mean or *kurtosis* can be calculated in SQL following the approach for the standard deviation. What do they mean? Skewness measures the symmetry of a distribution. A distribution is *symmetrical* if the mean is equal to the median. In that case, the mean is in the 'middle' of the distribution; when drawn, the distribution looks symmetrical around the mean. A symmetrical distribution has zero skew; but when a distribution is not symmetrical, it either has positive skew (or right skew, or right tail) or negative skew (or left skew, or right tail). The kurtosis describes the *tail* of a distribution, that is, values that are not close to the mean. In this sense, kurtosis is useful in that it indicates the propensity of a distribution to produce extreme values (i.e. outliers).[21] To compute skewness, we can use

```
SELECT sum(pow(Value - mean, 3)) / count(*)  /
       pow(sum(pow(Value - mean, 2)) / (count(*) - 1), 1 / 3)
FROM Data, (SELECT avg(Attr) as mean FROM Data) AS T;
```

Recall that the formula `pow(x, 1 / 3)` is used to express $\sqrt[3]{x}$ as $x^{\frac{1}{3}}$. However, in many systems this formula is not numerically stable, so results should be checked for accuracy.

To calculate kurtosis, we can use

```
SELECT sum(pow(Value - mean, 4)) / count(*)  /
       pow(sum(pow(Value - mean, 2)) / (count(*) - 1), 1 / 3)
FROM Data, (SELECT avg(Attr) as mean FROM Data) AS T;
```

As before, one must be aware that this formula may not be numerically stable.

We turn our attention now to categorical attributes. The most important tool in this case is the *histogram*. A histogram, in its simplest form, is a list of values in the domain together with their *frequency* (number of times the value appears). We have already seen how to calculate histograms earlier, when computing the mode. The general schema to build a histogram of categorical attribute `A` is

```
SELECT A, count(*)
FROM R
GROUP BY A;
```

[21] Because the kurtosis of a univariate, normal distribution is always 3, the kurtosis of a dataset can be computed and compared to this value: if it is greater than 3, it indicates that the dataset may have outliers. However, this is just a heuristic.

For instance, suppose that we have a demographic dataset with an attribute `zip code`; we can count how many residents live on each zip code with

```
SELECT zip-code, count(*) as population
FROM Dataset
GROUP BY zip-code;
```

Histograms are really one example of a more general technique, *binning* (also called *bucketing*). In binning, the values of a variable are divided into disjoint intervals (called *bins* or *buckets*), and all the values that fall within a given interval are replaced by some representative value. This makes the technique applicable to continuous (numerical) variables too. For categorical values, it is customary that each value is its own interval, but this does not necessarily have to be the case. Binning also generalizes histograms in that the representative value for an interval is not limited to the frequency; it can be another statistic too.

We can do general binning in SQL. For this, the values must be distributed into intervals, so the first task is to determine the number of intervals. There are basically two approaches to this. The first one is called an *equi-depth* histogram, and it sets the boundaries of the bins so that each has an equal number of data points. This is usually done by defining *quantiles*, cut points that split a domain into intervals with the property that each interval has as many data points as any other (plus or minus one, if the number of data points does not divide the number of intervals). The most commonly used quantiles are

- the *percentiles*, which divide the domain into 100 intervals, so that each one has 1% of the data;
- the *quartiles*, which divide the domain into 4 intervals, corresponding to 25, 50, 75, and 100% of the data;
- the *deciles*, which divide the domain into 10 intervals, corresponding to 10, 20,..., and 100% of the data.

The median is sometimes called a 2-quantile, since it splits the data into two intervals, both with equal number of points too.

Example: Binning with Quartiles

Assume a table `Heights(age,size)` where the age is an integer, and we want to create a histogram for the `age` attribute, and we want our histogram to be based on quartiles, the first bin representing the first 25% of all ages, the second one representing from 25 to 50%, the third one from 50 to 75%, and the fourth one above 75%. First, we need to find out how many elements there are and divide them into 4 groups. Second, we want to divide all the elements into those 4 groups; this is done after sorting them.

```
WITH OrderedData AS
(SELECT *, number
 FROM Heights,
```

```
      (SELECT count(*)/4 as number FROM Heights) AS T
  ORDER BY age)
SELECT 'bottom25' as quartile, age, size
FROM OrderedData
LIMIT number
UNION
SELECT '25to50' as quartile, age, size
FROM OrderedData
LIMIT number OFFSET number
UNION
SELECT '50to75' as quartile, age, size
FROM OrderedData
LIMIT number OFFSET number*2
UNION
SELECT 'top25' as quartile, age, size
FROM OrderedData
LIMIT number OFFSET number*3;
```

The table that results from answering this query can in turn be used to calculate other values, like $IQS(X)$, the inter-quantile range of the sample (the number of values in the middle 50% of the data).

Clearly, this approach is too cumbersome to be used with more bins (imagine the example with percentiles!). Modern versions of SQL have tools to make this task much simpler; they are discussed in Sect. 5.3.

The second approach is the *equi-space* histogram, which makes all bins the same *width* (i.e. all intervals of the same size; this approach is sometimes called *fixed-width*). In categorical attributes, this means the same number of values on each interval but, unless there is a special reason to group certain categorical values together, this approach is normally only used for ordinal and numerical domains. Note that the width of the bins determines, for a given numerical domain D, the number of bins; if h is the chosen width, then the number of bins is $\lceil \frac{max(D)-min(D)}{h} \rceil$ bins.

Example: Fixed Width Histograms

Assume table `Heights(age,size)` as before. We again want to create a histogram for the `age` attribute, and we want our intervals to have a fixed width, say 4.

```
SELECT age, size, ceil(((age-minage)+1)/4) as bin
FROM Heights, (SELECT min(age) as minage FROM Heights) as T
ORDER BY bin;
```

This maps each value to a bin starting at 1; the minimal age gets mapped to 1 by (age-minage)+1; the second smallest to 2, and so on. Note that the ceiling of the division acts as a modulus function, since this time we are using integer division. With the result of this query, we can group by bin and manipulate the data as required.

In general, if we have a column with attribute `Values` on it and want an equi-space histogram of `Values` with width n,

```
SELECT Values, ceil(((Values - minval)+1)/n) as bin
FROM Data, (SELECT min(Values) as minval FROM Data) as T
ORDER BY bin;
```

will do the trick.

Exercise 3.31 To avoid outlier problems, one can try getting rid of extreme values. Repeat the equi-spaced (fixed-width) histogram of `Heights` after removing the largest and smallest age. Hint: this removal should be one prior to computing anything for constructing the histogram.

One has to be careful here as divisions into bins that are too ‘wide’ (large intervals) result in a few, coarse bins and may hide important characteristics of the data, while a divisions into bins that are too ‘narrow’ or ‘thin’ results in a large number of bins, which may make the data look quite irregular if it does not fit well into any known distribution. There is no general rule to choose the number of bins, but there are several rules of thumb. The simplest one is pick, for n data points, $\sqrt{n}$ intervals for the histogram. There is also *Sturges' rule*: the number of bins for n data points should be (assuming all bins have equal width) $\lceil(\log_2 n + 1)\rceil$, but this works best for normal distributions of moderate size. Another approximation is to use the sample's standard deviation, s, and calculate $\frac{3.5s}{n^{\frac{1}{3}}}$. Instead of $3.5s$, another possible value is $2IQS(X)$, where $IQS(X)$ is the inter-quantile range of the sample (which we just saw how to calculate). All these methods are very sensitive to *outliers*, which force the bins to become too wide. One may want to ignore outliers in these calculations, by removing extreme values prior to breaking up the data into bins.

Sometimes the bins are determined by the semantics of the attribute being analyzed. As we have already seen, if cut points can be established with domain knowledge, we can bin based on them.

Example: General Binning

Assume a dataset with a `price` attribute; we are interested in determining how many ‘cheap,’ ‘medium,’ and ‘expensive’ products we have, having determined beforehand what those are.

```
SELECT type, count(*)
FROM Data
GROUP BY (CASE WHEN price > 100 THEN 'expensive'
          WHEN price <= 100 and price > 50 THEN 'medium'
          ELSE 'cheap' END) as type;
```

Exercise 3.32 In `ny-flights` dataset, count the number of flights that are ‘very late’ (arrival delay >20 min), ‘late’ (arrival delay between 10 and 20 min), 'somewhat late' (arrival delay <10 min), 'on time' (no arrival delay), and 'early' (negative arrival delay).

Exercise 3.33 Using the previous exercise as a temporary result (hint: use a `WITH` clause or a `FROM` subquery), count the number of flights of each type for each airline.

Exercise 3.34 In the `world` database, count the number of countries that call themselves a 'Republic' and the ones that do not (attribute `GovernmentForm`).

Note that for each bin, only a count is kept, so other information about the distribution (like mean) may be hard to recover. One can keep additional information if needed or desired (for instance, the mean of each bin). From the histogram, we can identify generic properties of the data distribution, like symmetry and skewness, as well as whether the distribution is unimodal, bimodal, or multimodal. This can help us narrow down the choice of a theoretical distribution that fits the data. Alternatively, we can compare the histogram generated from the data with the histogram that a theoretical distribution would generate, as described in Sect. 3.2.3. Histograms have their limitations. In particular, distributions with heavy tails are usually not well accounted for with histograms.

A couple of variations of binning can be useful. Sometimes we may want to use percentages instead of raw counts; this is sometimes called a *normalized* binning or normalized histogram.

Exercise 3.35 Using the example shown, compute a normalized, equi-space histogram for the fictitious `Heights` table.

Another idea, especially useful with ordinal attributes, is to have *cumulative totals* on each bin: if the bins can be put in a certain order, each bin counts its own values plus all the values of the bins preceding it. This idea can be applied to any type of histogram.

Example: Cumulative Histogram

Assume a simple histogram of heights `HHeight` that we want to use to calculate cumulative counts.

```
WITH HHeight(value, freq) AS
 (SELECT height AS value, count(*) as freq FROM Height)
SELECT D2.value, sum(D1.freq) as cumulative
FROM HHeight D1, HHeight D2
WHERE D1.value <= D2.value
GROUP BY D2.value;
```

The query joins the `HHeight` table with itself, but the condition `D1.value <= D2.value` matches each tuple in copy `D2` of the table with all those tuples in copy `D1` that have smaller values; when grouping by `D2.value`, all the frequencies of those smaller values are added up.

The same idea can be used to compute cumulative percentages, or other types of binning.

Example: Cumulative Binning

Assume we have split the table `Heights` into bins numbered 1,2,...,10 by calculating deciles; the result is in table `Deciles(decil,frequency)`. We now want a cumulative histogram.

```
SELECT D1.decil, count(*)
FROM Deciles D1, Deciles D2
WHERE D2.decil <= D1.decil
GROUP BY D1.decil;
```

This query takes a join of table `Deciles` with itself (hence the renaming), but on the D2 side it puts all values of `decil` that are less than or equal to ('precede') the current decil we are considering for grouping. That is, when `D1.decil` is 3, the join qualifies deciles 1, 2, 3 for `D2.decil`; this is what is counted by `count(*)`.

Exercise 3.36 Compute a cumulative histogram of percentages over table `HHeight`.

As we will see, using window functions (Sect. 5.3) makes some of these tasks, like calculating quantiles and cumulative results, much easier.

We close this section by presenting a powerful idea that is the basis of some sophisticated analysis. As we have seen, it is possible to associate with any type of attribute (categorical, ordinal, or numerical) a probability distribution by counting the frequency of each value and normalizing all such counts with the total number of data points. Once this is done, we can calculate the *entropy* of an attribute A, which is traditionally defined as

$$H(A) = \Sigma_{a \in A} P(a) log P(a)$$

that is, we add the product of each probability with the logarithm (usually, in base 2, although other bases can be used) of the probability. In SQL,

```
SELECT sum(Pa * log(Pa))
FROM (SELECT A, sum(1.0 /total) as Pa
      FROM Data, (SELECT count(*) as total FROM Data) AS T
      GROUP BY A);
```

The idea of entropy is to represent the 'information content' of attribute A, in the following sense: the 'information content' of value $a \in A$ is considered inversely proportional to its probability of happening; since common or normal events are expected to happen, their occurrence is not very informative. In contrast, uncommon, rare events are unexpected, and therefore their occurrence is very informative. The entropy of A is the 'average' of the information content of all values of A. This simple definition is the basis of very sophisticated forms of analysis; we will see a couple of the most elementary ones in the next section.

Exercise 3.37 Calculate the entropy of attribute `GNP` in the `Country` table of the `World` database.

3.2.2 *Multivariate Analysis*

Here we look at possible connections between attributes. We focus on the case of two single attributes; investigating sets of attributes is much more complex (some of it is done under full data analysis).

Let A and B be two attributes from our data table. There are two ways of looking at relationships between A and B: in one, we can assume that one of them is *caused* or influenced by the other. In this case, we say there is an *independent* attribute (in statistics, a *predictor*) and a *dependent* attribute (in statistics, an *outcome* or *criterion*). In the other case, we can assume that they do not depend on each other, although they may still be connected (for instance, there can be a third attribute Z that causes both A and B, therefore linking their values).

In either case, analysis can be further subdivided depending on the type of attributes we are dealing with. We can distinguish 3 cases: both attributes are categorical; both are numerical; and the mixed case (one attribute categorical, one numerical).[22]

Most of the time, we may not know for sure whether two attributes are independent or not. Hence, we start analysis with some simple tests to try to determine whether there is some connection between the attributes. The simplest test, and one that can be used in all types of attributes, uses their probabilities (recall that we saw early on how to calculate the probabilities of categorical attributes using grouping and counting; the same approach can be used with ordinal attributes). Given attributes A and B, we compute the probability of A, $P(A)$, and the probability of B, $P(B)$, as usual; we also compute the *joint probability* of A and B, $P(A, B)$:

```
SELECT A, B, sum(1.0/total) as JointProb
FROM Data, (SELECT count(*) as total FROM Data)
GROUP BY A, B;
```

We can create, for a given table `Data` with attributes A and B, another table `Probabilities` that has $P(A)$, $P(B)$, and $P(A, B)$ as new attributes. Once this is one, we use the following test: if $P(A, B) = P(A)P(B)$, the attributes are *independent*. This can be done very easily: all we have to do is check whether the following query returns zero:[23]

[22] Ordinal attributes can many times be grouped with numerical for the purposes of this subsection.

[23] Recall that, due to numerical instability, we may see a very small, but non-zero, result.

```
WITH ProbA AS
  (SELECT A, sum(1.0/total) as PrA
   FROM Data, (SELECT count(*) as total FROM Data) as T
   GROUP BY A),
      ProbB AS
  (SELECT B, sum(1.0/total) as PrB
   FROM Data, (SELECT count(*) as total FROM Data) as T
   GROUP BY B),
      ProbAB AS
  (SELECT A, B, sum(1.0/total) as PrAB
   FROM Data, (SELECT count(*) as total FROM Data) as T
   GROUP BY A, B)
SELECT sum(PrAB - (PrA * PrB))
FROM (SELECT A, B, PrA, PrB, PrAB
      FROM ProbA, ProbB, ProbAB
      WHERE ProbA.A = ProbAB.A and ProbB.B = ProbAB.B)
      AS Probabilities;
```

Note that we must compute the single and joint probabilities apart, since they require different groupings (different ways to look at the data).[24] Note also that we compute probabilities for all values of A and B, but when joining with the joint probability, only combinations that occur in the data are kept.

Probabilities are also used in other important measures, like *Pointwise Mutual Information (PMI)*, defined as follows:

$$PMI(A, B) = \log \frac{P(A, B)}{P(A)P(B)}.$$

Obviously, it is also the case that when this measure is zero, attributes are independent. In SQL:

```
SELECT log(sum (PrAB / (PrA * PrB)))
FROM Probabilities;
```

The *mutual information* of A and B (in symbols, $I(A, B)$) in turn exploits PMI to give a measure of how dependent A is on B and vice versa:

$$I(A, B) = E_{P(A,B)} PMI(A, B),$$

where E is the expectation (mean) calculated over $P(A, B)$.

Exercise 3.38 Assuming a table `Probabilities` as above, with $P(A)$, $P(B)$, and $P(A, B)$, write an SQL query to compute $I(A, B)$.

The closer the mutual information is to zero, the more independent the attributes are. Because this value is not normalized, it is hard to interpret on its own; mutual information is often used to compare pairs of attributes for feature selection or other tasks.

[24] In Statistics, the single probabilities are called, in this context, *marginal* probabilities.

For numerical attributes, the *covariance* is the simplest measure of a possible connection between attributes. Covariance is given by

$$C(A, B) = E((A - E(A))(B - E(B))),$$

where A, B are the attributes and E the expectation (mean). In a sample, this resolves to:

$$C(A, B) = \frac{\Sigma_{i=1}^{N}(a_i - \bar{A})(b_i - \bar{B})}{N-1}(1) = \frac{\Sigma_{i=1}^{N} a_i b_i}{N} - \frac{(\Sigma_{i=1}^{N} a_i)(\Sigma_{i=1}^{N} b_i)}{N(N-1)}(2),$$

where $\bar{A}$ denotes the mean (average) of A and $\bar{B}$ the mean (average) of B.

The covariance can also be expressed via the formula:

$$Cov(A, B) = E(A, B) - E(A)E(B),$$

which is easier to compute if necessary. Note the similarity to the test of independence using the joint probability distribution; just like the mean of an attribute is related to its probability distribution, the covariance is related to the join probability. This is why it is another test for independence.

In PostgreSQL, there is an aggregate function, `covar_samp(X,Y)`, to compute the covariance of two attributes in a table. There is no aggregate for covariance in MySQL as of this writing (2019), but it is easy to simulate from the definition above. The simplest way to write the covariance of attributes A and B is to use the last definition:

```
SELECT (avg(A*B) - (avg(A)* avg(B)))
FROM Data;
```

Exercise 3.39 Write an SQL query to compute the covariance using the original formula (1).

Exercise 3.40 Write an SQL query to compute the covariance using the original formula (2).

Exercise 3.41 Compute covariance not from original data but from table `Probabilities` with $P(A)$, $P(B)$, and $P(A, B)$.

The Pearson correlation coefficient is simply a normalized covariance:

$$\rho = \frac{Cov(A, B)}{std(A)std(B)},$$

where $std(A)$ is the standard deviation of A, and similarly for $std(B)$. This value is always between -1 and $+1$, with larger absolute values reflecting stronger correlation, and the sign indicating positive (both A and B move in the same direction) or negative (A and B move in opposite directions) correlation. If two attributes are independent, the value should be zero. However, if this value is zero, it does not mean that the attributes are independent (there could be a non-linear relationship), because correlation has an important drawback: it only detects *linear* associations between the variables. As a simple (and typical) example, let $A = (-2, -1, 0, 1, 2)$ and $B = (4, 1, 0, 1, 4)$. Then $B = A^2$, but their correlation is zero.[25]

In PostgreSQL, the aggregate `corr(A,B)` computes the correlation of attributes A and B. There is no aggregate for correlation in MySQL; fortunately, this is another concept that can be expressed in SQL—in several ways, actually. The simplest way to expression correlation in SQL is

```
SELECT (avg(A*B) - (avg(A)* avg(B))) / (std(A) * std(B))
FROM Data;
```

when average and standard deviation are available; or, if an aggregate for covariance exists:

```
SELECT Covar(A,B) / (std(A) * std(B))
FROM Data;
```

Exercise 3.42 A formula typically used for correlation is

$$r_{AB} = \frac{n\Sigma(a_i b_i) - (\Sigma a_i \Sigma b_i)}{\sqrt{n\Sigma a_i^2 - (\Sigma a_i)^2}\sqrt{n\Sigma b_i^2 - (\Sigma b_i)^2}}.$$

Write the SQL query that implements that formula.

Just like covariance was connected to probability, so is Pearson correlation, so it can be computed from probabilities too.

Exercise 3.43 Compute Pearson correlation not from original data but from table `Probabilities` with $P(A)$, $P(B)$, and $P(A, B)$.

For ordinal attributes, it is common to use *rank correlation*, a measure of the relationship between the rankings on each variable. The idea here is to compare the rank of the attributes (their position in the order), instead of the attribute values. We assume that our table `Data` contains, besides attributes A and B, attributes $Arank$ and $Brank$ giving their ranks in their respective orders (i.e. they both are $1, 2, \ldots, n$ for a dataset with n rows). There are several rank correlation measures; the most popular ones are Kendall's τ and Spearman's ρ. The idea behind Kendall's is as follows: given two pairs of (X, Y) values (x_1, y_1) and (x_2, y_2), we say they are *concordant* if the ranks of both elements agree: either the rank of x_1 is higher than that of x_2 *and* the rank of y_1 is higher than that of y_2 or the rank of x_1 is lower than

[25](Pointwise) mutual information could be used here as is not limited to linear associations, so it is a more general measure of dependence, but it is also harder to interpret.

that of x_2 *and* the rank of y_1 is also lower than that of y_2. Otherwise, we say the pairs are *discordant*. Kendall's rank correlation is computed as

$$\tau = \frac{2(\text{number of concordant pairs})(\text{number of discordant pairs})}{n(n-1)},$$

where n is the number of data pairs. To calculate this in SQL, we need to compare the values of $Arank$ and $Brank$:

```
SELECT
 sum(CASE WHEN ((D1.Arank < D2.Arank AND D1.Brank < D2.Brank)
             OR (D1.Arank > D2.Arank AND D1.Brank > D2.Brank))
           THEN 1 ELSE 0 END) as concordant -
 sum(CASE WHEN ((D1.Arank < D2.Arank AND D1.Brank > D2.Brank)
             OR (D1.Arank > D2.Arank AND D1.Brank < D2.Brank))
           THEN 0 ELSE 1 END) as discordant
     / (count(*) * (count(*) - 1)
FROM Data D1, Data D2;
```

Note that we are using the Cartesian product of the dataset with itself, in order to consider all possible pairs of data points. Note also that when either of the ranks coincide, the pair is neither concordant nor discordant, so we must check for both conditions explicitly.

The value of Kendall's correlation is always between -1 and $+1$: if the agreement between rankings is perfect, we get $+1$; if the disagreement is perfect, we get -1; if the ranks are independent, we get 0.

Exercise 3.44 The above formula for Kendall is expensive due to the Cartesian product. A more direct way to calculate this correlation is

$$\frac{2}{n(n-1)} \Sigma_i (sign(Arank_i - Brank_i),$$

where $Arank_i$ is the rank of the ith element in A, $Brank_i$ is the rank of the ith element in B, and the function `sign` simply tells us whether its argument is positive, zero, or negative and is available in many SQL systems (including PostgreSQL and MySQL). Using this, express this definition in SQL.

To calculate Spearman's Rho, we can use the definition directly:

$$\rho = 1 - \frac{6\Sigma_i d_i^2}{n^3 - n},$$

where $d_i = Arank_i - Brank_i$ is the difference in rank between the ith pair of elements, as above.

```
SELECT 1 - (6 * sum(pow(Arank - Brank, 2))) /
            (pow(count(*), 3) - count(*))
FROM Data;
```

When one attribute is categorical and another continuous, there are two ways to consider influence: in one, the continuous variable influences the categorical one. The typical approach here is to see if we can use the continuous variable to predict the continuous one using *classification*, which is explained in the next chapter. On the other direction, the typical approach is to see if the categorical attribute has an influence on the continuous value by analyzing the differences between the means of the values generated by each category. In full generality, this is the ANOVA (Analysis of Variance) approach from statistics. When talking about a single categorical variable and a single continuous one, we can use a simplified version, usually called *one-way ANOVA*. We explain this technique through an example:[26] assume we have a table `Growth(fertilizer, height)` where we keep the results of an experiment: several plants where grown for a period of time using different kinds of fertilizer. The first attribute gives the class of fertilizer used, and the second one gives the growth of a plant using that type of fertilizer. We take the following steps:

1. Calculate mean of continuous value for each type:

```
CREATE TABLE group-means AS
SELECT fertilizer, avg(height) as group-mean
FROM Growth
GROUP BY fertilizer;
```

2. Calculate overall mean of the group means.

```
SELECT avg(group-mean) as overall-mean
FROM group-means;
```

3. Calculate the sum of square differences between group mean and overall mean; divide over number of groups minus 1 (degrees of freedom). Note that this is the *between-groups* variability.

```
SELECT sum(pow(group-mean - overall-mean, 2)) * size /
       (num-groups - 1) AS between-groups
FROM (SELECT count(*) AS size FROM Growth),
     (SELECT count(DISTINCT fertilizer) as num-groups
      FROM Growth),
     group-means;
```

4. Calculate the sum of square differences between the group mean and the values on each group and divide by number of groups times number of data points minus 1. Note that this is the *within-groups* variability.

```
SELECT sum(pow(height - group-mean, 2)) /
       (num-groups * (size - 1)) AS within-groups
FROM Growth, group-means
WHERE Growth.fertilizer = group-means.fertilizer;
```

[26]The example is taken from Wikipedia but changed to reflect the structure of the data as a tidy table.

5. The F-ratio is the within-groups variability divided by the between-groups variability. This ratio can be compared to the *F-distribution*, for which tables exist, to determine if the value obtained is significant (this is sometimes called an *F-test*).

To put it all together in a single query, we use the typical strategy of pre-computing needed results with the WITH clause and subqueries in the FROM clause:

```
WITH group-means AS
     SELECT fertilizer, avg(height) as group-mean
     FROM Growth
     GROUP BY fertilizer
SELECT between-groups / within-groups
FROM (SELECT sum(pow(group-mean - overall-mean, 2)) *
             (size * (num-groups - 1)) AS between-groups
      FROM (SELECT count(*) AS size FROM Growth),
           (SELECT count(DISTINCT fertilizer) as num-groups
            FROM Growth),
           group-means) AS temp1,
      (SELECT sum(pow(height - group-mean, 2)) /
               (num-groups * (size - 1)) AS within-groups
       FROM Growth, group-means
       WHERE Growth.fertilizer = group-means.fertilizer)
           AS temp2;
```

When both attributes are categorical, we use the basic histogram technique to compare counts, in what is called a *contingency table* or *cross-tabulation*. Based on the counts, we can compute the *chi-square* (χ^2) *test* of independence. In its simplest form, the idea is to compare the number of co-occurrences one observes between their values with the number of co-occurrences one would expect if the attributes were independent of each other. This is a basic technique that can also be applied to ordinal and numerical attributes by counting their frequencies.

The idea is simple: if attribute A has n possible values and attribute B has m possible values, let c_{ij} be the count of data with value i of A and value j of B, $c_{i\_}$ the count of data with value i of A, $c_{\_j}$ the count of data with value j of B (these last two are sometimes called *margin sums*), and c the total count of values; then we call $E(i, j) = \frac{c_{i\_} c_{\_j}}{c}$ the *expected* value of i, j and compare this with c_{ij}, the *observed* value (since it comes from the data):

$$T(X, Y) = \Sigma_i \Sigma_j \frac{(c_{ij} - E(i, j))^2}{E(ij)}.$$

This is the χ^2 test of independence. The data is usually presented as a matrix (spreadsheet) with the margin sums, with format:

	Y_1	Y_m	Row total
X_1	c_{11}	c_{1m}	$c_{1\_} = \Sigma_j c_{1j}$
$\vdots$	$\dots$	$\dots$	$\vdots$
X_n	c_{n1}	c_{nm}	$c_{n\_} = \Sigma_j c_{nj}$
Column total	$c\_1 = \Sigma_i c_{i1}$	$c\_m = \Sigma_i c_{im}$	n

where $c_{i\_} = \Sigma_j c_{ij}, c\_j = \Sigma_i c_{ij}$ (this two-way table of A-B counts is an example of bivariate histogram).

As we saw in Chap. 2, in databases the data should be in a tidy table; this means an attribute describing the rows, another one describing the columns, and another one giving the values for each combination—that is, we would have a table with schema `Data(X,Y,Attr)`. Getting the counts from this is very easy; for instance, the different $c_{i\_}$ are given by

```
SELECT X, count(*)
FROM Data
GROUP BY X;
```

and likewise for the B counts c_j and the joint counts c_{ij}. The result of the tabulation is compared to the chi-square distribution with $(m-1) \times (n-1)$ degrees of freedom; if the result if greater than the chi-square, then the attributes are not independent. If the result is smaller than the chi-square, then the attributes are independent.

Example: Chi-Square Test

Assume a table that gives, for the subscribers to a magazine, the city where they reside and their status—whether they are currently active subscribers or have stopped their subscription.[27] The question is whether the city where they live affects the rate of subscription.

CUSTOMERS		
City	Status	Count
Gotham	Active	1462
Gotham	Stopped	794
Metropolis	Active	749
Metropolis	Stopped	385
Smallville	Active	527
Smallville	Stopped	139

[27] This example comes from a website that, unfortunately, seems to be invisible to searches right now.

To answer the question, we proceed as follows:

1. Compute the margins: total by status, by city, and total overall.
2. Compute the overall stop (active) rate: total stop (active) divided by total.
3. Compute the expected values (per city): total number per city times overall rate (for active and passive).
4. Compute the deviation: the difference between observed and expected value (for active and passive).
5. Compute the χ^2: deviation squared divided by expected value.

In SQL, this can be written in a single query, using WITH and subqueries in the FROM clause to organize the margin sum calculations:

```
WITH
 (SELECT Cust.city, Cust.status, Cust.Count,
         perCity * (perStatus / total) as expected
  FROM Cust,
        (SELECT city, sum(Count) as perCity
         FROM Cust
         GROUP BY city) as Cities,
        (SELECT status, sum(Count) as perStatus
         FROM Cust
         GROUP BY status) as Statuses,
        (SELECT sum(Count) as total FROM Cust)
  WHERE Cust.city = Cities.city
         and Cust.status = Statuses.status)
SELECT sum(pow(Count-expected,2)/expected);
```

Exercise 3.45 Assume a car company is testing 3 new car models, called A, B, and C. It tests the car for 4 days, giving a score between 1 and 10 each day. They end up with a matrix (spreadsheet):

	A	B	C
Day 1	8	9	7
Day 2	7.5	8.5	7
Day 3	6	7	8
Day 4	7	6	5

The question is: are there significant differences between the models? Do the test results depend on the day and not the model? Put the data in a relational format and carry out a χ^2 test.

Exercise 3.46 Assume the simplest contingency table, one with two binary attributes A and B: A can only take values x_1 and x_2, and B can only take values y_1 and y_2. Let us give names to the counts as follows:

		Y		
		y_1	y_2	
X	x_1	a	b	a+b
	x_2	c	d	c+d
		a+c	b+d	n

where $n = a + b + c + d$. Then we can write the chi-square test as follows:

$$\chi^2 = \frac{n(ad - bc)^2}{(a+b)(c+d)(a+c)(b+d)}.$$

Create a tidy table (one with schema `(X, Y, count)`) to represent the table above and write an SQL query to implement this formula.

Exercise 3.47 Using this contingency table again, we can define the *odds ratio* of A as $\frac{ad}{bc}$. Write an SQL query over your tidy table to compute this result.

Exercise 3.48 Again using this contingency table, we can define the *relative risk* of x_1 as $\frac{a/(a+c)}{b/(b+d)}$. Write an SQL query over your tidy table to compute this result.

3.2.3 Distribution Fitting

Sometimes, it is suspected that some of the (numerical) data in our dataset has been generated by a process that follows some standard probability distribution (at least approximately, since real data always has some noise). Whether this is the case, it can be checked by generating artificial values with a formula for the distribution and comparing what we obtain with the data values. When the difference is not significant, we consider that indeed the data has been generated by a process with an underlying distribution. This process is called *distribution fitting*. When used to check whether some hypothesis a researcher has actually holds in the data, this is part of the process of *(statistical) hypothesis testing*.

To guess a distribution, one can start by using histograms and calculating some basic measures of centrality and distribution, as indicated earlier. There are more sophisticated techniques, like Kernel Density Estimation, which we will not cover here. However one comes up with the initial guess, one should check that indeed the chosen distribution is a good fit. Several checks are available for this; chi-square can be used.

The general approach is this:

1. Set the data in the form of a distribution, that is, create a table with schema `(value, probability)`. This can be done as seen previously, by estimating probabilities from percentages, and percentages from raw counts of data.
2. Find out the parameters of this empirical distribution, in particular, mean and standard deviation.
3. Add an attribute for the estimate to the data table. We end up with a table with schema `(value, probability, estimate)`. This new third column has nulls on all rows for now.
4. Choose a distribution. Using the formula for this distribution, and the mean and standard deviation of the sample, generates an estimate for each value and leaves it on the `estimate` attribute.
5. Compare the estimate obtained (`estimate`) with the empirical `probability`.

Example: Fitting a Normal Distribution

A phone company records the lengths of telephone calls.[28] The data is converted into a table DATA with columns *number of minutes, observed number of calls with those minutes*. This can be done for each number of minutes (1,2,3,...) or as a histogram with time intervals, after choosing a width (0 to 2 min, 2 to 4 min, etc. for a width of 2 min). We calculate estimates for μ (mean) and σ^2 (variance) from the data in table DATA. We then apply the density function of the normal distribution:

$$f(x) = \frac{1}{\sqrt{2\pi\sigma^2}} e^{-\frac{(x-\mu)^2}{2\sigma^2}}$$

to the *number of minutes* column (so *number of minutes* is x) to get the column *estimated number of minutes*. In SQL:

1. Create table DATA and populate it with data. If the data is in *raw* format `(callid, number of minutes)`, a simple group by and count query will produce the table DATA with probabilities instead:

```
CREATE TABLE DATA AS
SELECT num-minutes, sum(1.0/ total) as prob
FROM Raw-data, (SELECT count(*) as total FROM Raw-data)
GROUP BY num-minutes;
```

2. Create a table with columns for the data, the observed frequencies, and the expected frequencies and fill in first two columns from the data and the last one with nulls.

```
CREATE TABLE NORMAL(data,observed-freq,expected-freq) AS
SELECT num-minutes, freq, null
FROM DATA;
```

[28] This example is from [11].

Note that the previous two steps can be combined into one; we separate them to explain the process in a step-by-step manner, but they can be easily combined (see Exercise below).

3. Calculate mean and standard deviation in the data as usual. Then populate the column for expected frequencies by applying the formula of the normal distribution to the data column and the mean and standard deviation. In PostgreSQL,

```
UPDATE NORMAL
SET expected-freq = (1 / sqrt(2*pi()*stddev)) *
    exp(- (pow(observed-freq - avg, 2) / (2*stdev)));
```

4. Compare observed frequencies and expected frequencies and decide if the difference is small enough. A very simple way to do this is to add the absolute differences and express them in terms of standard deviations (see Exercises). A more sophisticated way to attack the problem is to run a chi-square test.

Exercise 3.49 Combine the first two steps above, creating the table NORMAL in one single query.

Exercise 3.50 Write an SQL query to compare observed and expected frequencies as suggested above (sum of absolute differences over standard deviation).

To fit a different distribution, we simply use the appropriate formula. For instance, we may want to fit a Poisson distribution to data. This distribution rises very rapidly, and once the peak is reached it drops very gradually. This is typical of discrete events, where the non-occurrence makes no sense. This is also typical of arrival and departure events, and it is why this distribution is so frequently used in queue theory. The density function for Poisson is

$$\frac{e^{-\mu}\mu^x}{x!},$$

where μ as usual is the mean of the distribution, $x!$ is the *factorial* of x, and x is the expected number of times the event can happen in a given time period (so the above can be calculated for $x = 1, 2, \ldots$). For instance, if x is the number of orders per day in an e-commerce site, this is the percentage of days we can expect x orders. Thus, we can build a table `Poisson(Xval, Expected)`, where for $Xval = 1, 2, \ldots$ we can calculate the expected value with the formula above, multiplying the result (which is a percentage) by a total value of x. For example, for 100 days, we multiply the percentage by 100, which will tell us how many days, out of 100, we can expect x orders. Again, this is an expected value. We can compare this with the observed value from the data, using χ^2 again. Note that the Poisson values can also be accumulated, to determine the days where orders will be x or less.

Exercise 3.51 A website collects the number of orders they get each day for a certain time period (n days). This is then converted to a table DATA with schema *number of orders per day, observed number of days with that number of orders*. From the original data or from DATA, we can calculate the mean, which is used for μ. Then the table DATA is expanded by using the density function and computing, for each number of orders per day, the Poisson expected value: that is, x is the number of orders per day, and we calculate the expected percentage of days that would have that value (column *expected percentage*). This expected percentage is then multiplied by the total number of days to get the expected number of days with the value (column *expected number of days*). Implement this in SQL with some made-up values.

Fitting distributions can be used to find and remove outliers, as we will see in Sect. 3.3.3.

3.3 Data Cleaning

Data cleaning is the set of activities and techniques that identify and fix problems with the data in a dataset. In general, we want to identify and get rid of 'bad' or 'wrong' data. However, it may be very hard to identify such data, depending on the domain and data type:

- For (finite) enumerated types, a simple membership test is needed. For instance, months of the year can be expressed as integers (in which case only values between 1 and 12 are allowed) or as strings (in which case only value "January," ..., "December" are allowed). For this types, distinguishing 'good' from 'bad' data is usually not hard, unless some non-standard encoding is used (for instance, using 'A,' 'B,' ... for months of year).
- For pattern-based domains (like telephone numbers or IP addresses), a test can rule out values that do not fit the pattern. However, when a value fails to follow the expected pattern, it may be a case of bad formatting; such values must be rewritten, not ignored, or deleted (see next section for more on this). Telling badly formatted from plain wrong values can become quite hard.
- In open-ended domains (as most measurements are), what is a 'bad' or 'wrong' value is not clear cut. An in-depth analysis of the existing values and domain information have to be combined to infer the range of possible and/or likely values, and even this knowledge may not be enough to tell 'good' from 'bad' values in many cases.

In general, the problems attacked at this stage may be quite diverse and have different causes; hence, data cleaning is a complex and messy activity. Different authors differ in what should be considered at this stage; however, there are some basic issues that most everyone agrees should be addressed at this point, including

- *Proper data*. For starters, we need to make sure that values are of the right kind and are in a proper format. As already mentioned in Sect. 2.4.1, when we load data into the database we must make sure that values are read correctly. In spite of our efforts, we may end up with numbers that are read as strings (because of commas in the values, or other problems), or dates that are not recognized as such and also read as plain strings. Even if the data is read correctly, the format may be not appropriate for analysis. Hence, making sure data is properly represented is usually a first step before any other tasks.

- *Missing values*. In the context of tabular data, this means records that are missing some attribute values, not missing records. The issues of whether the data records we have in a dataset are all those that we should have and whether the dataset is a representative sample of the underlying population are very important but different and treated in detail in Statistics. Missing values refers exclusively to records that are present in the dataset but are not complete—in other words, missing *attribute* values. Detection may be tricky when the absence of a value is marked in an ad-hoc manner in a dataset, but the real problem here is what to do with the incomplete records. The general strategies are to ignore (delete) the affected data (either the attribute or the record), or to predict (also called *impute*) the absent values from other values of the same attribute or from value of other attributes. The exact approach, as we will see, depends heavily on the context.
- *Outliers*. Outliers are data values that have characteristics that are very different from the characteristics of most other data values in the same attribute. Detecting outliers is extremely tricky because this is a vague, context-dependent concept: it is often unclear whether an outlier is the result of an error or problem, or a legitimate but extreme value. For instance, consider a person dataset with attribute `height`: when measured in feet, usually the value is in the 4.5 to 6.5 range; anyone below or above is considered very short or very tall. But certainly there are people in the world who are very short (and very tall). And, of course, we would expect different heights if we are measuring a random group of people or basketball players. In a sense, an outlier is a value that is not *normal*, but what constitutes normal is difficult to pin down [15]. Thus, the challenge with outliers is finding (defining) them. Once located, they can be treated like missing values.
- *Duplicate data*. It is assumed that each data record in a dataset refers to a *different* entity, event, or observation in the real world. However, sometimes we may have a dataset where two different records are actually about the same entity (or event, or observation), hence being duplicates of each other. In many situations, this is considered undesirable as it may bias the data. Thus, detecting duplicates and getting rid of them can be considered a way to improve the quality of the data. Unfortunately, this is another very difficult problem, since most of the time all we have to work with is the dataset itself. The simpler case where two records have exactly the same values for all attributes is easy to detect, but in many cases duplicate records may contain attributes with similar, but not exactly equal, values due to a variety of causes: rounding errors, limits in precision in measurement, etc. This situation make duplicate detection much harder. The special case of *data integration* (also called *data fusion*), where two or more datasets must be combined to create a single dataset, usually brings the problem of duplicate detection in its most difficult form [6, 7].

There are other activities, like discretization, that are sometimes covered under data cleaning. In this textbook, some of those activities are discussed in the next section under the heading of *data pre-processing*. Here, we focus on the four topics introduced above. However, in real life one may have to combine all activities in a feedback loop; for instance, one may have to standardize values before analyzing them looking for outliers or duplicates.

3.3.1 Attribute Transformation

In general, *attribute transformations* are operations that transform the values of an attribute to a certain format or range. These are needed to make sure that data values are understandable to database functions, so that data can be manipulated in meaningful ways. Many times it may be necessary to transform data before any other analysis, even EDA, can begin.

For categorical attributes, a typical transformation is *normalization standardization*, a process whereby each value is given a unique representation so that equality comparisons will work correctly. For instance, a comparison between strings may differentiate between uppercase and lowercase characters, something that may create artificial distinctions (i.e. if values like 'Germany' and 'germany' are present in the dataset). For numerical attributes, typical transformations include *scaling* and *normalization*. Because these transformations are highly dependent on the data type, they are discussed separately next.

3.3.1.1 Working with Numbers

Two common operations on numerical values are *scaling* and *normalization*. Scaling makes sure that all values are within a certain range. *Linear scaling* transforms all values to a number between 0 and 1; the usual approach is to identify maximum and minimum values in the domain (or in the attribute, as present in the database[29]) and transform value v to $\frac{v - min}{max - min}$. This is easily implemented:

```
UPDATE Dataset
SET Attr = (Attr - (SELECT avg(Attr) FROM Dataset)) /
            ((SELECT max(Attr) FROM Dataset) -
             (SELECT min(Attr) FROM Dataset));
```

Unfortunately, this will not work in all SQL databases (for instance, it works in Postgres, but it does not in MySQL), due to the following: we are changing values in table `Dataset`, but we also want some statistics (mean, minimum, maximum) obtained from it. The intended meaning, of course, is that such statistics must be

[29]What statisticians call the sample.

obtained from the table *before* any changes are applied to the table—and that is indeed what Postgres does. However, some systems will not understand this order of evaluation and will consider that we are asking to examine and modify table `Dataset` at the same time. To work around this problem, sometimes the query must be rewritten to isolate the statistics computation and make clear that it should happen before any changes:

```
UPDATE Dataset
SET Attr = (Attr - (SELECT mean
                    FROM (SELECT  avg(Attr) as mean
                          FROM Dataset) as D1)) /
                   (SELECT maxa - mina
                    FROM (SELECT max(Attr) as maxa,
                                 min(Attr) as mina
                          FROM Dataset) as D2 ) ;
```

Another scaling that provides values between 0 and 1 but is not linear in nature is the *logistic* scaling where value v is mapped to

$$\frac{1}{1+e^{-v}}.$$

The logistic function is used when the presence of outliers (very large or very small values) is suspected, as it can accommodate them at the top or the bottom of the range. This function gives the typical 'sigmoid' graph, softly approaching the minimum and maximum (0 and 1) without ever reaching it.

Exercise 3.52 Assume a generic table `Dataset` with attribute `Attr`. Implement logistic scaling of `Attr` in SQL.

Normalization consists of making sure that all values are expressed using a known quantity as the unit. The typical example is the *z-score*, where values are expressed in terms of standard deviations from the mean:

$$Z(v) = \frac{x - \mu}{sdev},$$

where μ is the mean of the domain of x and $sdev$ the standard deviation. This again is easy to express in SQL:

```
UPDATE Data
SET Attr = (Attr - (SELECT avg(Attr) FROM Data)) /
           (SELECT std(Attr) FROM Data);
```

Many other transformations are possible.

Exercise 3.53 The same caveat as in the previous transformation applies: in some systems, like MySQL, this query needs to be written with subqueries to force an evaluation order. Do this following the model provided in the scaling case.

When we have an attribute with a skewed distribution (see Sect. 3.2.3) needs to be corrected, the variable is typically transformed by a function that has a disproportionate effect on the tails of the distribution. The most often used transformations in this case change value v using the log transform ($log(v)$ in base 2, e or 10, commonly), the multiplicative inverse ($\frac{1}{v}$), the square root ($sqrt(v)$), or power ($power(x, n)$ for some n).

3.3.1.2 Working with Strings

In general, cleaning categorical attributes means making sure that one and only one name (string) represents each category. One does not want small, irrelevant differences (like letters being uppercase or lowercase) to interfere with tasks like grouping or searching for values.

Because strings can be used to represent many diverse values, there are no hard and fast rules about how to deal with strings. Ideally, one would like to make sure that the strings being used are standardized, that is, they use a set of standard names so that no confusion can occur. This is easy in the case of closed (enumerated) domains; for others, the best approach is to come up with a set of *conventions* that are followed through the analysis.

Example: String Normalization

Assume a student table that contains, among other attributes, the name of the department where the student is majoring. Unfortunately, and because of manual entry, this attribute contains many different ways of spelling the same value:

Student-id	...	Department
1	...	Dept of Economics
2	...	Econ
3	...	Department of Econ

All these names should be unified to a single, canonical one. In this simple example, the following SQL command will do the trick:

```
UPDATE Student
SET Department = "Economics"
WHERE position('Econ' IN Department) > 0;
```

As we will see, `position` is a string function that tells us at what position the string denoted by the first argument appears in the second; asking for a position greater than zero simply means that the string 'Econ' has been found in the value of attribute `Department` in a given tuple. In this case, this happens to capture all variations of the name. In more complex, real-life cases, several commands may be needed to capture all cases.

The above example is typical of how string functions are used to put string values in an appropriate format. The most common tasks are to 'trim' the strings (getting rid of whitespaces and other non-essential characters), to modify the string to some 'standard' form (like making sure the whole string is lowercase), or to extract part of a string for further analysis.

There are many string functions, and unfortunately different systems may express them in slightly different ways. Instead of trying to go over each possibility, we organize the functions by what they do and give some examples of each type, focusing on commonly used functions that are present in both Postgres and MySQL:

- Functions that *clean* the string: they change it by getting rid of certain characters or transforming existing characters. Among them are TRIM(), LOWER(), and UPPER(). The inverse of this (adding characters to a string) is called *padding*, expressed with LPAD(), RPAD(), and others.
- Functions that *find* elements (characters or substrings) within a given string. The most popular ones are POSITION() and STRPOS().
- Functions that *extract* elements of a string or *split* a string into parts. This includes functions SUBSTR() and SPLIT(). The inverse of this (putting together several strings into a single one) is usually called *concatenation*, expressed by CONCAT().

Other functions typically available include several forms of *replacement*, where parts of a string are removed and other characters substituted for them, like SUBSTR().

Finally, most functions include the useful function LENGTH(), which returns the number of characters in a string.

We now provide examples for some of these functions. A typical cleaning function, `TRIM(position characters FROM string)`, removes any character in `characters` (the default is whitespace) from string `string` starting at `position` 'leading' (leftmost), 'trailing' (rightmost), or 'both' (the default). Postgres also has functions `LTRIM()` (equivalent to `TRIM('leading,'...)`) and `RTRIM()` (equivalent to `TRIM('trailing,'...)`). As an example of use, in the dataset `ny-rolling-sales`, several attributes (like Neighborhood) have a fixed length, meaning they take the same space regardless of actual length. This may result in padding (extra blanks added to the value); to get rid of it, we write

```
SELECT TRIM(both ' ' FROM Neighborhood)
FROM ny-rolling-sales;
```

to get the values without padding (note that whitespaces in the middle are not affected).

LOWER() is used to force every character in a string to become lowercase (if the character is already in lowercase, it is left untouched). A similar effect is achieved with UPPER().

```
SELECT UPPER(address)
FROM ny-rolling-sales;
```

These functions are usually applied to categorical attributes in combination with UPDATE so that values are described in a uniform manner; this way, searches do not miss values and grouping works correctly.

Among finding functions, `POSITION(substring IN string)` (equivalently, `STRPOS(string, substring)`) returns a numerical value denoting the character position where the substring first appears in the string (character positions are numbered starting at 1 on the left; if the substring is not found, 0 is returned). This function is useful because the value it returns can be used by other functions, as we will see.

```
SELECT POSITION('East' IN address)
FROM ny-rolling-sales;
```

Extracting functions include `LEFT(string,` *n*`)`, which gets *n* characters from the beginning (left) of the string (function `RIGHT(string,` *n*`)` does the same from the end of the string). This function is useful when all string values follow a certain pattern and we need to extract a part of the string based on those patterns.

Example: Combining String Functions

Old datasets from the Imdb website[30] contain a field where both movie title and year are combined together as the value of a (single) attribute, as in "Amarcord (1973)"; we can extract the year with

```
SELECT RIGHT(title, 6)
FROM imdb;
```

since the year (together with the parenthesis) constitutes the last 6 characters of each title. If we wanted to extract the title, we cannot rely on a fixed position, but we know we should go all the way to the opening parenthesis, '(,' so we could write

```
SELECT LEFT(title, STRPOS(title, '('))
FROM imdb;
```

The number of characters to extract is calculated by finding the position where the '(' appears (in our example, 9). Note that, when evaluating functions, just like in Math, innermost functions are evaluated first.

To extract a part of a string starting somewhere in the middle, we use `SUBSTR(string, start-pos, length)`, which extracts the substring of its first argument that starts at the second argument and has as many characters as the

[30]The Internet Movie DataBase, https://www.imdb.com.

third argument indicates. Thus, another way to extract the year (this time without parenthesis) is to use

```
SELECT SUBSTR(title, 9, 4)
FROM imdb;
```

If we want to combine string values instead of splitting them, we can use the CONCAT function. This function simply takes a sequence of strings and returns a single string that is the combination of all the arguments. The arguments can be a mixture of attribute values and constants; this can be used to put a value in a certain format. For instance, to separate title from year using a hyphen, we could use

```
SELECT CONCAT(LEFT(title, STRPOS(title, '(')), ' - ',
              SUBSTR(title, 9, 4))
FROM imdb;
```

Most systems allow the use of the two pipe characters (||) as synonym for concatenation.

Exercise 3.54 Consider the `Name` attribute in table `city` of database `world`.

1. Some values in this attribute include a second name in parenthesis. Display `Name` without such second names.
2. Some values in this attribute are compound; but the parts of the name are separated by hyphen (-), sometimes not. Display `Name` with all hyphens suppressed.
3. Display compound names always with hyphen (i.e. change whitespaces between words to hyphens).

Finally, many systems include a useful function that allows concatenating string values *across rows*. Because they work in sets of rows, technically these functions are aggregates, but they behave unlike typical aggregate functions, which are numerical. The string concatenation version (called `GROUP_CONCAT()` in MySQL and `STRING_AGG()` in Postgres) simply creates a new value by putting all its arguments together; usually, a character separator is used (comma in MySQL; specified as the second argument, in Postgres) and sometimes it is possible to specify an order. As any other aggregate, this one can be used with grouping.

Example: String Concatenation

Recall table `Country` in database `world`, where each row contains information of a certain country, including attributes `Continent`, `name`, and `Population`. Then the MySQL query:

```
SELECT Continent,
       GROUP_CONCAT(name ORDER BY Population DESC)
FROM Country
GROUP BY Continent;
```

will return one row per continent, and for each one a single string listing all country names, separated by comma, and ordered by their population. The result will look like

```
+-----------+-----------------------------------------+
| Continent | GROUP_CONCAT(Name ORDER BY Population DESC)|
+-----------+-----------------------------------------+
| Asia      | China,India,Indonesia,Pakistan,Bangladesh,.|
| Europe    | Russian Federation,Germany,United Kingdom,.|
| North America | United States,Mexico,Canada,Guatemala,.|
| Africa    | Nigeria,Egypt,Ethiopia,Congo, The Democr.. |
| Oceania   | Australia,Papua New Guinea,New Zealand,Fiji|
| Antarctica| Heard Island and McDonald Islands,Antarcti.|
| South America |Brazil,Colombia,Argentina,Peru,Venezu...|
+-----------+-----------------------------------------+
```

3.3.1.3 Working with Dates

Dates are some of the most problematic types of data in SQL. This is due to the fact that dates can be expressed in many different formats: months can be expressed by name (January, etc.) or by number; years are sometimes written in full (2019) or shorted by century (19); the order of elements can change (year-month-day, month-day-year, day-month-year,...). In many countries, the standard format is DD-MM-YYYY (that is, two digits for the day first, followed by two digits for the month, followed by four digits for the year), but in the United States, the format is MM-DD-YYYY (month goes first). The SQL standard form for DATE is defined as YYYY-MM-DD. The reason to adapt this format is that, if sorted, it yields chronological order.[31] The standard for TIME is HH:MM:SS[.NNNNNNN] (two digits for hours first, followed by two for minutes, followed by two digits for seconds, and optionally seven digits for fractions of a second).[32] The SQL standard for DATETIME (timestamps) is a DATE plus a TIME, separated by a space: YYYY-MM-DD HH:MM:SS. The standards for INTERVAL distinguish between two classes of intervals:

- the YEAR TO MONTH class, with format: YYYY-MM.
- the DAY TO FRACTION class, with format: DD HH:MM:SS.F.

In Postgres, DATE, TIME, and TIMESTAMP follow the standard. For instance, '2017-01-08 04:05:06' is a valid timestamp. Other formats are also allowed; examples of valid dates are: 'January 8, 2017'; '2017-Jan-08'; '08-Jan-2017'. The date '1/8/2017' is allowed, but note that it is ambiguous. Examples of valid times

[31]This format is also in agreement with the ISO 8601 standard.

[32]This format is almost the ISO 8601 standard; it differs in the fractional seconds.

are: '04:05:06,' '04:05,' '040506,' '04:05 AM.' Note that any date or time value needs to be enclosed in single quotes.

Interval values are written using the pattern:
`quantity unit [quantity unit...] [direction]`,
where `quantity` is a number (possibly signed); `unit` is one of: microsecond, millisecond, second, minute, hour, day, week, month, year, decade, century, millennium, or abbreviations or plurals of these units; `direction` can be "ago" or nothing at all. For instance, `1-2` denotes the interval 1 year 2 months; `3 4:05:06` denotes 3 days 4 h 5 min and 6 s.

Most systems support *arithmetic* on time data: '+' can be used to add two dates, or two intervals, or add an interval to a date:

- Date '2001-09-28' + integer '7' results in date '2001-10-05' by adding seven days to the first argument.
- Date '2001-09-28' + interval '1 h' results in timestamp '2001-09-28 01:00:00' by adding 1 h to the original date, assumed to denote midnight of that day.
- Date '2001-09-28' + time '03:00' results in timestamp '2001-09-28 03:00:00' by adding 3 h to the original date, again assumed to denote midnight.
- Interval '1 day' + interval '1 h' results in an interval '1 day 01:00:00,' that is, one day and 1 h.

Likewise, '−' can be used for subtraction.

The function EXTRACT is used to get information bits from a date or timestamp:

```
SELECT cleaned_date,
       EXTRACT('year'   FROM cleaned_date) AS year,
       EXTRACT('month'  FROM cleaned_date) AS month,
       EXTRACT('day'    FROM cleaned_date) AS day,
       EXTRACT('hour'   FROM cleaned_date) AS hour,
       EXTRACT('minute' FROM cleaned_date) AS minute,
       EXTRACT('second' FROM cleaned_date) AS second,
       EXTRACT('decade' FROM cleaned_date) AS decade,
       EXTRACT('dow'    FROM cleaned_date) AS day_of_week
  FROM Data;
```

Other convenient functions to manipulate temporal data include:

- `current_date()` returns the current date.
- `current_time()` returns the current time of the day.
- `now()` returns the current date and time.
- `make_date(year int, month int, day int)` creates a date value.
- `make_interval(years int DEFAULT 0, months int DEFAULT 0, weeks int DEFAULT 0, days int DEFAULT 0, hours int DEFAULT 0, mins int DEFAULT 0, secs double precision DEFAULT 0.0)` creates an interval.
- `make_time(hour int, min int, sec double precision)` creates a time.
- `make_timestamp(year int, month int, day int, hour int, min int, sec double precision)` creates a timestamp.

Casting can also be used to create times. In fact, going from strings to dates often involves some text manipulation, followed by a *cast*, an operation where the system is told to coerce a value into a certain type. This is indicated by '::' in Postgres, as the next example shows.

Example: Casting Strings into Dates

Assume the data contains a string-based attribute `thedate` with values like '15-04-2015' and we want to convert them into dates:

```
SELECT (SUBSTR(thedate, 7, 4) || '-' || LEFT(thedate, 2) ||
        '-' || SUBSTR(thedate, 4, 2))::date AS date
  FROM Data;
```

The string functions break down the original string value into parts (the year starts at position 7, and it takes 4 characters; the month starts at position 4 and goes for 2 characters; and the day are the first (leftmost) 2 characters) that the casting "::date" can convert into a date attribute.

Also, functions `TO_DATE(text, pattern)` and `TO_TIMESTAMP(text, pattern)` can be used for conversion. The pattern argument indicates how the conversion should interpret the string; for instance, `to_date('05 Dec 2000', 'DD Mon YYYY')` indicates that '05' is the day, 'Dec' the value of the month, and '2000' the value of the year.

In addition to these functions, the SQL OVERLAPS operator on intervals is supported in most systems:

- `(start1, end1) OVERLAPS (start2, end2)` is true if `start1` is earlier than `start2` but `end1` is later than `start2` but earlier than `end2`, or vice versa;
- `(start1, length1) OVERLAPS (start2, length2)` is true as before, with `end1 = start1 + lenght1` and `end2 = start2 + length2`.

MySQL is a bit more restrictive about temporal values than Postgres is. DATE values must have the standard format of year-month-day (for example, '98-09-04'). However, the year may be 2 or 4 digits (with two digits, values 70 to 79 are given to the twentieth century, and values 00 to 69 to the 21st). Also, when expressing the date as a string, one can use any punctuation character as delimiter (so '98/09/04' and '98@09@04' are also okay), and even no delimiters ('980904') as far as the system can make sense of the value as a date. It is also possible to express the value as a number (again, as far as the system can make sense of the value as a date). Finally, MySQL has a "zero" value of '0000-00-00' as a "dummy date," which can be used in place of NULL for missing values. The same is true of DATETIME and TIMESTAMP values: they should follow the standard, but they can be expressed as a string with whatever delimiter the user chooses (or no delimiter).

As for functions, MySQL uses `ADDDATE()` or `DATE_ADD()` to add a time value to a date, and `DATE_DIFF()` to subtract a date from another. Function `ADDTIME()` adds two times, and `DATE_SUB()` subtracts a time from a date.

Example: Date Functions in MySQL

Here is an example that uses date functions. The following query selects all rows with a date_col value from within the last 30 days:

```
mysql> SELECT something FROM tbl_name
    -> WHERE DATE_SUB(CURDATE(),INTERVAL 30 DAY) <= date_col;
```

An important function is `DATE_FORMAT(date,format)`, which formats the date value according to the format string. The format is expressed by using a sequence of *specifiers*, characters preceded by the percent (%) sign, which indicate how to lay out the date. For instance, specifier `%a` is used for weekday name, abbreviated (Sun..Sat); `%W` for weekday name, not abbreviated (Sunday . . . Saturday), and `%w` for weekday as a number (0=Sunday..6=Saturday). Similar choices exist for month, year, hour, minutes, and seconds.

Formatting Dates in MySQL

The query

```
SELECT DATE_FORMAT("2019-08-12", "%M %d %Y");
```

returns

```
August 12 2019
```

Specifier `%M` indicates that the month is shown in full name; `%d` that they day is shown as a number; and `Y` that the year is shown as a 4 digit number.

Creating Times in MySQL

When creating a table for the `ny-flights` dataset, times like arrival and departure were modeled as integers, since this is how the raw values were expressed ('547'). However, this is not convenient for meaningfully manipulating the values (for instance, finding differences). To transform such values into times, we can use the function `MAKE_TIME(h,m,s)`, where argument `h` is an integer giving the number of hours; argument `m` is an integer giving the number of minutes; and argument `s` is an integer giving the number of seconds. To transform '547' into time '5:47:00,' we can use

```
SELECT MAKE_TIME(arr_time DIV 100, arr_time MOD 100, 0)
FROM ny-flights;
```

A similar approach will work for other times in the dataset.

Exercise 3.55 Transform scheduled departure time and departure time attributes in `ny-flights` as shown and take their difference (as times); compare this difference with the attribute `dep_delay`. Do you get the same values?

Exercise 3.56 Combine attributes `year`, `month`, `day`, `hour`, and `minute` into a time (note that you will need a 'seconds' value) and compare the result to attribute `time-hour`. Do you get the same values? What seems to be the difference?

3.3.2 Missing Data

The first problem with missing values is identifying them. In some dataset, missing values are not explicitly identified: the value is simply not present. In csv files, for instance, a missing value is identified by a value not being in its place; this results in two commas being adjacent (',,') or a record (line) ending in a comma instead of a value. In some datasets, explicit markers are used; unfortunately, there are no 'standard' or 'typical' markers, so the dataset must be examined carefully for such markers. However, some approaches are common. When there is an identifiable range of values, it is common to use values that are outside the range to mark missing data, so that they are not confused with regular values. For instance, a -1 may be used in a numeric field that is supposed to have only positive values. For strings with delimiters (quotes), sometimes the *empty string* ("") is used. In those cases, it is necessary to eliminate those 'special' values: the -1 would confuse many statistical analyses, and the empty string would still be treated as a string by a lot of software. Another typical value in categorical fields is the strings 'n/a' (or its variations 'N/A' or 'NA'), which stand for 'not available.' Since inside the database there is only one legitimate way to identify missing data (with the *Null* marker), we must identify and substitute such values when data is loaded into the database (see Sect. 2.4.1) or right afterward, before any analysis.

We assume in the rest of this section that missing values have been located in the data and substituted by Null markers, if not already identified by them. This can be achieved with the UPDATE command; for instance, imagine a dataset *Patients* with information about patients in a medical study. One of the attributes is `Height`, and another one is `Weight`, representing, respectively, the height and weight of each patient at the beginning of a study. We load it into table `Patient`, and note that some Heights and Weights are missing, indicated by a −1. We then 'clean' the table as follows:

```
UPDATE Patients
SET Weight = NULL
WHERE Weight = -1;
```

and similarly for Height.

Recall that SQL's Null is not a value, but an indicator that there is a 'hole' in a tuple, i.e. a Null denotes the *absence* of a value. Because of this, Nulls behave in somewhat idiosyncratic ways. For the purposes of Data Analysis, the most important characteristics of Nulls are: how they interact with comparisons in the WHERE clause, with aggregates in the SELECT clause, and with grouping when a GROUP BY is present.

With respect to conditions, the most important characteristic to remember is that all comparisons with Null fail. Technically, comparisons in SQL can return, besides a True and False result, and 'Unknown' result, and this is what happens when a value is compared with a Null—but since queries only return tuples that yield a True result in comparisons, we can think of comparisons with Null as returning False. As an example, assume that we are looking, in the Patients table, at patients who may be overweight, and we put a cut point of 220 lbs.[33] The query

```
SELECT *
FROM Patient
WHERE Weight > 220;
```

should retrieve all such patients. However, imagine that (for whatever reasons) we do not have the weight for some patients; the attribute Weight has some Nulls on it. All such tuples will not be retrieved, since the condition Weight > 220 will return 'Unknown' on those tuples, and the system only uses, for answering a query, tuples that return 'True' to conditions in the WHERE clause.

The fact that comparisons with Nulls always fail extends to even comparisons of two Nulls (they also fail) and to calculations with Nulls. Assume, for instance, that we are trying to approximate a BMI (Body Mass Index) calculation; this can be done by dividing weight by the square of height (when using the metric system, with weight in kilograms and height in meters; a corrective factor is applied when using pounds and inches). We could write a query like (assuming metric measurements for simplicity)

```
SELECT *
FROM Patient
WHERE Weight > power(Height, 2) * 18.5;
```

with 18.5 considered the threshold of a normal BMI. However, if either Weight or Height or both have Nulls, this calculation will fail: the comparison with

[33] That is approximately 100 kgs.

$>$ will return 'Unknown' because both sides of the $>$ (expressions `Weight` and `power(Height, 2) * 18.5`) are Null, and even `Null = Null` returns 'Unknown.' This means that we could be missing patients with a condition but with unknown weight or height.

This behavior of Nulls extends to more complex predicates as follows:

- Conjunctions (AND) are true when both conditions used are true. Since a comparison with Null is not true, a conjunction involving a comparison with Nulls will fail, regardless of what other comparisons do. As an example, the query

  ```
  SELECT *
  FROM Patient
  WHERE Weight > 220 and Height < 6;
  ```

  will fail on tuples where `Weight` is Null, regardless of what `Height` is. The whole comparison will also fail if both `Weight` and `Height` are Null or if only `Height` is Null.
- Disjunctions (OR) are true when at least one of the conditions used is true. When one of the conditions used involved Nulls, the whole comparison depends on what the other one returns. As an example, the query

  ```
  SELECT *
  FROM Patient
  WHERE Weight > 220 or Height < 6;
  ```

  will return tuples where the `Height` attribute is indeed less than 6 feet, even if `Weight` is Null. But even an OR will fail if both `Height` and `Weight` are Nulls. This behavior leads to some strange situations; the query

  ```
  SELECT *
  FROM Patient
  WHERE Weight > 220 or Weight >= 220;
  ```

  would seem to return every tuple of table `Patient`, since the condition is a tautology (whatever the weight of a patient is, it surely is greater than 220 or less than or equal to 220). However, tuples where the weight is Null will *not* be returned. This is important if we are *splitting* the dataset into classes or subsets using a predicate like this one.
- Negation (NOT) flips the result of a comparison: True becomes False, and False becomes true. However, negation leaves 'Unknown' unchanged, since we do not know what to flip. As a result, if the first example above were written as

  ```
  SELECT *
  FROM Patient
  WHERE NOT (Weight <= 220);
  ```

  the query would still fail to retrieve tuples where the attribute `Weight` is null.

The fact that all predicates fail with Nulls leads to an interesting question: what to do if we want to actually find those tuples where we have Nulls? SQL provides two special predicates, IS NULL and IS NOT NULL, that do exactly that. Either one of them takes an attribute name and returns True only for those tuples where the attribute has a Null marker. That is, the query

```
SELECT count(*)
FROM Patient
WHERE Weight IS NULL;
```

will tell us how many missing values there are in attribute Weight. Conversely,

```
SELECT count(*)
FROM Patient
WHERE Weight IS NOT NULL;
```

will tell us how many real (non-missing) values there are in attribute Weight. Using IS NOT NULL is akin to eliminating all rows of data where missing information is present.[34]

With respect to aggregates, we must be aware of the following:

- As a rule, *aggregates ignore Nulls*, with the exceptions below. As a result, if we use an aggregate in a query, like

  ```
  SELECT Sum(Weight)
  FROM Patient;
  ```

  the result will be the sum of all available weights, with nulls ignored. This seems commonsensical for Sum, but it is also the case for the other basic aggregates—Count, Avg, Min, and Max.
- An exception to the above is Count(*). This aggregate essentially counts rows, without concern as to the contents of the rows. In particular, whether attribute values are present or are Null is irrelevant for this aggregate.
- If *all* values in an attribute are Null, aggregates return a *default* value: this is 0 (zero) for count, but it is Null for all other aggregates (sum, avg, min, and max).
- When operating in the context of a GROUP BY, the behavior is the same: within each group, aggregates ignore Nulls, with the exceptions noted—but keep on reading to see how GROUP BY itself handles Nulls.

With respect to grouping, Nulls behave differently than they do with comparisons. Recall that in comparisons, Nulls are not equal to each other. Thus, if we use an attribute A in a GROUP BY clause, and A contains one or more Nulls, it would seem that each one of them would generate its own, separate group. However, GROUP BY treats Nulls *as if they were the same*: all Nulls in an attribute generate a unique group. Suppose, for instance, that we are trying to generate a histogram of

[34] For those familiar with R, this is similar to the na.rm = TRUE optional argument in R, and to the na.omit predicate in data frames (but the SQL IS NOT NULL only operates on single attributes, not whole tuples).

weight values, and we write

```
SELECT Weight, count(*)
FROM Patient
GROUP BY Weight;
```

If attribute `Weight` contains *one or more* Nulls, one single separate group will be created. That group will contribute one tuple to the answer: `(Null, n)`, where `n` is the number of tuples with Nulls in `Weight`. If we write

```
SELECT Weight, avg(Height)
FROM Patient
GROUP BY Weight;
```

and attribute `Weight` contains *one or more* Nulls, we will again see a tuple `(Null, n)` in the result, where `n` is the average height of all tuples with Nulls in `Weight` (if `Height` itself has Nulls, they will be ignored when computing the averages).

Because the behavior of SQL is somewhat inconsistent, and because we want our data to be 'clean' for analysis, it is customary to try to get rid of nulls. The choices are to eliminate nulls or to substitute a value for them. In the first case, we can choose to delete from the dataset the rows where some null is present, or to eliminate from the dataset the attribute where nulls are present. Once nulls are detected and we know which attribute (or attributes) are at fault, both operations are quite easy in SQL—the first one calls for a command like

```
DELETE FROM (Table)
WHERE (attribute) IS NULL;
```

The second one calls for a command like

```
UPDATE TABLE
DROP ATTRIBUTE (attribute);
```

Note that, in both cases, we likely lose data with the nulls. If the proportion of nulls is very small, dropping rows (observations) may be the best way to proceed, but dropping the attribute could result in a substantial loss of data.

However, determining values to substitute for nulls is much more complex; it involves *imputing* or *predicting* the missing value from some other data. How to do this depends on the kind of missing value we are dealing with.

Several authors [4, 16] distinguish 3 types of missing values: let A be an attribute where some values are missing, and B be all other attributes in the table with A. Values missing in A can be

- *missing completely at random (MCR)* or *observed at random*: the values are missing independently of the underlying value of A, and of any values of B. In this case, the missing values can be filled in because the values of A that are present give us a good idea of the distribution of all values in A, so we can infer the underlying distribution and replace the missing values by the mean of the present values or by fitting a distribution to see which value of A is likely to be under-represented.

- *missing at random (MAR)*: the values are missing independently of the underlying value of A but may depend on other values B. This implies that that values of A are correlated with the value of (some of) the attributes in B. In this case, the missing values can be filled if we can find out what attribute(s) in B is (are) related to A, and the nature of the relationship. In such cases, we can apply a `predictor` method to the values of the relevant attribute(s) in B on rows where the value of A is missing. Two typical methods to use are *linear regression* and `k-nearest neighbors`, both explained in the next chapter.
- *non-ignorable missing values*: the values are missing independently of other values B but may depend on the underlying value of A. The problem with this case is that there may not be a way to replace the missing values meaningfully, since the values of other attributes do not help, and there is a connection with other values of A, which means that those other values of A that we have are not an impartial guide to the missing values, as in the first case.

An example from [4] makes the difference clear: assume a sensor for air temperature that runs out of battery at some point in time. The battery running dry is not related to the air temperature, or to any other weather variable, so we are in the first case. Here, we would be justified in replacing the missing values with the mean temperature, since the mean of the present values is (under the assumption that the missing values are random) an unbiased estimator of what the missing temperature is. Now assume a person is in charge of replacing batteries in the air sensor, but he or she does not do it when it rains. Then the battery is more likely to be dead when it is raining (one of the B attributes), although it has nothing to do with air temperature. This is the second case. Here, if we find out the connection between rain and temperature, we can try to infer an appropriate temperature from the value of the attribute 'rain.' We can, for instance, take the mean temperature of the raining days only or apply linear regression to the rain attribute in order to predict temperature (see next chapter). We would be justified in doing this because of the relationships between the attributes. Finally, assume that the sensor malfunctions at temperatures under zero degrees. Then other variables may be not related (i.e. it may rain or not rain at any temperature), but the sensor fails in a manner related to the missing value. In this case, imputing a value is very problematic: clearly, we cannot use the other temperatures, since they represent a whole range of possible temperatures, while the missing values come from a small subset of the range (below zero temperatures). If there is no connection with other attributes, they do not yield enough information to impute a value. Finally, note that even if we are aware of the situation (we know for a fact that there is a sensor malfunction at low temperatures) and we can infer that the missing values are low values, we still do not have enough information to determine a good value for each missing one.

How to tell the cases apart? One way is to try to determine whether there is some connection between A and some other attributes in the table. A first approach is to identify an attribute B that may be connected to A; using correlation, as explained in the previous section, is a first step (although, as we have mentioned, correlation only detects *linear* relationships). Using PMI may also be a good idea, as PMI is more general. However, in both cases we do not obtain a direct answer to the question "Is B related to A (and hence a decent predictor)?" When B is numerical, a preliminary test can be to compare values of B when A is absent to values of B when A is present:

```
SELECT avg(B) as mean,
       CASE (WHEN A IS NULL THEN 'absent'
             ELSE 'present') AS Avalue
FROM Data
GROUP BY Avalue;
```

If the means are sufficiently different, we can deduce that there is a connection. If no connection is found with any attribute, we can rule out the second case and we are in either the first or third one. Unfortunately, it may not be possible to distinguish between these two from the data alone.

Note that much simpler methods are always available; for instance, in a numerical attribute, it may be tempting to get rid of nulls by using zeroes as this allows arithmetic to proceed without problems (in some systems, trying to divide by a null may cause an error). However, such approaches may introduce bias, in that it may disguise the true distribution of the attribute or its relationships to other attributes. Hence, this is not a desirable approach and should be avoided—and, if used, it should be properly documented in case it needs to be undone.

Finally, assume that have decided we are going to get rid of nulls by substituting them with some new value. To see what the attribute would look like, we can use

```
SELECT CASE(WHEN attribute IS NULL THEN (newvalue)
            ELSE attribute END) AS new-attribute
FROM Data;
```

However, there is a simpler way to achieve this using the `COALESCE` function:

```
SELECT COALESCE(attribute, newvalue) AS new-attribute
FROM Data;
```

The `COALESCE` function accepts a list of values and returns the first one that is not null. In this case, it will simply return the value of `attribute` when it does not contain a null, or `newvalue` otherwise, and so it accomplishes the same as the CASE construction in a more succinct manner.

3.3.3 Outlier Detection

An outlier is a value that is not 'normal,' in the sense that it is quite different from other values in a domain. It tends to be *extreme* (for numerical attributes) and *unusual* (for categorical attributes). While the notion is highly intuitive, there is no formal definition of what it means for a data point to be an *outlier*, since this depends on the context. Outliers may indicate a data quality problem (an error in data acquisition/representation), or they may indicate a random fluctuation, or they may represent a truly infrequent, exceptional, or abnormal situation. In some applications, outliers are exactly what we are looking for; as an example, in credit card fraud, the fraudulent transactions are the outliers among a sea of legitimate transactions. In other applications, outliers are bad values that disturb the general nature of the data, and it is a good idea to get rid of them. Reusing an example from the previous section, assume we receive data from a temperature sensor. The average temperature so far is 75 degrees Fahrenheit, and the standard deviation is 5 degrees. Suddenly we receive a reading of 100. Is this the result of a heat wave, or a faulty sensor? There is usually no way to tell just from the data, and the difference is crucial for addressing the issue: a bad reading should be deleted and replaced by another value (or simply considered missing and ignored); an extreme, but legitimate value, should be kept—it is a crucial piece of information.

Outlier detection for single attributes depends on the type of attribute: for nominal attributes, we can calculate frequencies for each possible value; an outlier will be a value with a very low frequency. For numerical values, it is harder to decide what is an outlier without assuming an underlying distribution. A possible tactic is to try and find a distribution that fits the data well (see Sect. 3.2.3). When a distribution fits the data reasonably except for a few values, those values can be considered outliers. This is commonly done with the standard distribution: in this distribution, 95% of values are within 2 standard deviations from the mean, and over 99% of all values are within 3 standard deviations, so values beyond that can be classified as outliers. However, in other distributions, for instance exponential distributions (or any distribution which is very skewed or has a long, heavy tail), it is not be possible to tell outliers from regular values with this test. To make things more complicated, note that when outliers are present in a dataset, they influence the very statistics we are using (i.e. they can move the mean, and they may make the standard deviation much larger). Thus, it may be a good idea to use more robust statistics, like the trimmed mean or the Median Absolute Deviation (MAD) introduced earlier, to search for outliers.

Finally, in some cases it may be necessary to do a full analysis like *clustering* to decide if a value is an outlier.

Example: Finding Outliers in Names

Assume a dataset about people where one of the attributes is last (family) name. We suspect some names may be not entered correctly (typos, mispronunciations). One way to check for this is to focus on very rare values (since errors tend to be different, each typo may generate different results): the query

```
SELECT last-name, count(*) as freq
FROM Dataset
GROUP BY last-name
HAVING freq = 1;
```

will show names that appear only once. This, by itself, does not make the values 'bad,' but it makes them deserving of some further scrutiny.

Example: Finding Outliers in Numbers

Instead of simply finding numbers that are n standard deviations from the mean, we can use the trimmed mean and trimmed standard deviation introduced in the previous section to more reliably identify outliers.

```
WITH TrimmedStats AS
 (SELECT avg(A) as tmean, stdev(A) as tdev
  FROM Data, (SELECT max(A) as Amax FROM Data),
             (SELECT min(A) as Amin FROM Data)
  WHERE A < Amax and A > Amin)
SELECT A
FROM Data
WHERE A < tmean - (2*tdev) or A > tmean + (2*tdev);
```

Exercise 3.57 Write an SQL query to find outliers by using MAD. A common rule is to consider outliers values that are more than $1.5x$ from the MAD value, where x is the number of standard deviations we consider significant [8].

We will see later in Sect. 5.3 more direct ways to use MAD to find outliers.

3.3.4 Duplicate Detection and Removal

Identifying duplicates, in tabular data, is the task of examining two or more rows and determining whether they refer to the same observation, object, or entity in the real world. This task is also known as *deduplication, entity linkage, data linkage, data/record matching, entity resolution, co-reference*, and *merge/purge*.

The hardest part of duplicate identification is to determine when two rows refer to the same object or entity. In the basic case, we expect one or more attributes to be identical. That is, we determine a set of attributes that could serve as primary key (perhaps expanded with additional attributes for caution) and check to see if two rows have the same values for those attributes. Note that a primary key, if created artificially, is useless for this purpose. For instance, in a people database, assume that gave an 'id' to each person entered but now suspect that there may be duplicates (the same person may have been entered more than once). To check that, we may focus on (first and last) name, address, and date of birth, reasoning that two people with the same name, living in the same address, and having the same birthday are a strong indication of a duplicate. The 'id,' if system generated, is of no use to determine duplication. In this case, a simple query will do:

```
SELECT fname, lname, address, dob, count(*)
FROM Data
GROUP BY fname, lname, address, dob
HAVING count(*) > 1;
```

However, in many cases this simple approach will not be enough. Many times, repeated records do not have the exact same values for the attributes, but very close or similar ones. The reasons may go from typos to measure noise to having incorrect data in the database. A more sophisticated method that can deal with small mistakes is *fuzzy matching* (also called *approximate matching*). The idea is that if two values are very close, we can consider them the same for the purposes of duplicate detection. Note that what is considered 'very close' depends on the context: on a date of birth attribute, one day apart is a lot, but on a 'shipping date' attribute, one day apart may be an error in data entering. Likewise in numerical attributes: in a measurement of distance between two cities, 1100 miles (or kilometers) is almost the same as 1101 miles (or kilometers), while in a measurement of screw length, 11 inches and 12 inches can be quite different.

Implementing the intuitive notion of 'close' is usually done by using the idea of *distance*, which we study in Sect. 4.3.1. Intuitively, a distance between two values is a number that expresses how 'far apart' (how different) they are: a small distance means the values are similar; a large distance, that they are dissimilar. For the case of numerical values, the distance one typically uses is the absolute difference, sometimes 'normalized' by some value. This can be expressed quite simply in SQL.

Example: Approximate Distance

Recall the `Patients` table; the query

```
SELECT *
FROM Patient,
     (SELECT max(height) as maxh FROM Patient)
WHERE abs(D1.height - 6.0)/ maxh < alpha;
```

will return all individuals in the dataset whose height is 'close' to 6.0 (six feet), where the idea of 'close' is represented by being, as a percentage of the largest height, less than some cut point `alpha`.

For nominal values, the idea of fuzzy matching is to consider two strings as similar if they have more similarities than differences. Recall that SQL provides a `LIKE` operator that compares a string to a pattern, but the options of this operator are limited (see Sect. 3.1 for a description of LIKE).

For the case of comparing two string values, most systems provide some functions to implement fuzzy matching. Postgres, for instance, has two methods in its `fuzzystrmatch` module:[35]

- Function `difference(string1, string2)` returns a number that expresses the differences between the *Soundex* of two strings. The Soundex system is a

[35] Additional modules are activated in Postgres with the command `create extension` *`module-name`*.

method of matching similar-sounding names. Because of the way English is pronounced, the same (or very close) sound may end up being written in very different ways, depending on the context.[36] Soundex attempts to undo this; `difference` compares the results of Soundex and gives a number between 0 (no similarity) and 4 (total similarity). A query like

```
SELECT last-name
FROM Dataset
WHERE difference(last-name, 'Jones') > 3;
```

will return all last names that would sound very similar to 'Jones' in English. Obviously, this method works only for English language names. MySQL provides a similar method `string1 SOUNDS LIKE string2`, which is a short for `SOUNDEX(string1) = SOUNDEX(string2)`, with `SOUNDEX()` a function that generates the Soundex of its input.
- Function `levenshtein(string1, string2)` calculates the *Levenshtein distance* between two strings. This distance computes the total number of letter changes that would be necessary to transform one string into the other. In this case, a higher number means more difference: zero means the strings are identical, and the number can be as large as the size (number of characters) of the larger string. This distance applies to any language.

```
SELECT last-name
FROM Dataset
WHERE levenshtein(last-name, 'Jones') < 2;
```

Unfortunately, MySQL does not provide a similar function.

Of note, Levenshtein is only one of several possible distances for strings; more sophisticated matches against text are explored in Sect. 4.5.

When comparing two data records in the dataset, we first need to determine which attributes are likely to identify each data record; for each attribute, choose some distance that we are going to use to decide when two values are 'close enough' and, finally, pick a method to combine all distances to decide when the data records are indeed the same or not. The typical approach is to 'add' all the individual attribute distances. This can done in several ways. Let $A_1, \ldots, A_n$ be the attributes being compared, and $dist_i$ the distance applied to values of attribute A_i. Then, two rows or records r and s are compared using $dist_1$ on $r.A_1$ and $s.A_1, \ldots, dist_n$ on $r.A_n$ and $s.A_n$. We can combine all these distances to obtain a total distance between r and s, $dist(r, s)$, in several ways:

- Directly: if distances are comparable across domains (for instance, if they are all numerical and based on normalized values, or normalized themselves), with

$$dist(r, s) = \frac{\Sigma_{k=1}^{n} dist_k(r.A_k, s.A_k)}{\Sigma_{k=1}^{n} max(dist_k)}.$$

[36] As many students of English as a second language have discovered to their consternation.

Note that the denominator is the largest possible distance between two items and serves to normalize the result. Note also that this only works if all distances are on the same scale; otherwise, they need to be normalized individually:

$$dist(r,s) = \Sigma_{k=1}^{n} \frac{dist_k(r.A_k, s.A_k)}{max(dist_k)}.$$

Usually, the result obtained is compared to some threshold.

- By *Boolean combination*: assume that we have a threshold for each distance measure to determine, on an attribute-by-attribute basis, whether two values are similar enough or not. Let δ_k be 0 if the similarity of values in attribute A_k is not 'good enough,' and 1 if it is. Then the formula

$$dist(r,s) = \frac{\Sigma_{k=1}^{n} \delta_k(r.A_k, s.A_k)}{n}$$

is the proportion of attributes where there is agreement (out of all n of them).

Other approaches are possible. For instance, either one of the two approaches introduced can be modified with *weights*: if not all attributes have the same importance, we can use a sequence $w_1, \ldots, w_n$ of weights to indicate the weight w_l to give to $dist_l(r.A_l, s.A_l)$. For instance, the direct approach would result in

$$dist(r,s) = \frac{\Sigma_{k=1}^{n} w_k dist_k(r.A_k, s.A_k)}{\Sigma_{k=1}^{n} max(dist_k)}.$$

Example: Duplicate Removal in Patient Dataset

In the `Patient` table, we decide that two patients are one and the same if their last names and weights are similar enough:

```
WITH Similarity AS
(SELECT D1.Id, D2.Id,
     ((levenshtein(D1.lname, D2.lname) / (maxl * 1.0)
     +
     abs(D1.height, D2.height) / range) as sim
 FROM Dataset D1,
     Dataset D2,
   (SELECT max(length(lname)) as maxl FROM Dataset) AS T1,
   (SELECT max(height) - min(height) as range FROM Dataset)
   AS T2)
SELECT D1.Id, D2.Id
FROM Similarity
WHERE sim > 0.9;
```

Exercise 3.58 Give the SQL query to determine whether two people are the same or not in the Patient dataset by using Boolean combination of distances on `last-name` and `weight`.

Exercise 3.59 Modify your SQL query to give a weight of 0.75 to `last-name` and 0.25 to `weight` (it is common to use weights that add up to 1).

Finally, another task related to duplicate detection is the detection of inconsistencies. Two records in the dataset are said to be inconsistent if they refer to the same entity/event/observation and have contradictory information. Assume, for instance, that in the people dataset we discover that there are two records for the same person, but each one of them gives a different age. Since a person can only have one age, we know we have an inconsistency here. Clearly, not both data points (ages) can be correct at the same time; thus, we should correct one of the two. Unfortunately, in many cases (as in this example) it is impossible to decide, just from the data, which value is correct. In fact, dealing with inconsistencies depends largely on having domain knowledge, and it can become quite difficult. For instance, if two records about the same patient are inconsistent with respect to weight, we know that this is a value that can change over time, so we could assume that we simply ended up with information about the same patient taken at different time points. In a case like this, we would like to keep the more recent data (since this makes the data more likely to be accurate; recall Sect. 1.4). However, this does not tell us which value is the more recent one.

One way to detect inconsistencies is to enforce as many rules about the domain as possible; see Sect. 5.5 for some guidance on how to do this.

3.4 Data Pre-processing

Even after cleaning, data may still not be ready for analysis. This is usually due to the fact that values, even if correct, are not in the format that analysis tools (or algorithms) assume them to be. Hence, additional operations are needed to prepare the data for analysis. Typical operations used at this stage include aggregation, sampling, dimensionality reduction, feature creation, discretization, and binarization [16]. We explain each one briefly:

- Aggregation: this consists of combining two or more data records into one. This is especially common when data can be seen at several *levels of granularity* or detail. For instance, data on some event that is taken at regular time intervals of 1 min can be seen as 'by-the-hour' by aggregating all records within 60 min. Aggregated data has less detail but tends to have lower variability, and overall patterns may be more clear. This is usually implemented using the GROUP BY operator in SQL.
- Sampling: choosing a subset of the data ('sample') to work on. This makes computation less expensive, so it is common to sample when we want to carry out

complex analysis over large datasets, or when trying several tentative analyses of the same data. A sample chosen at random is expected to be representative of the whole dataset, hence it can help to establish the properties of the dataset. The tricky part is making sure that a good procedure (one where each data record has equal chance of being chosen) is followed. In databases, the task is sometimes accomplished by simply picking some rows of a table with the help of some pseudo-random number generator, as this provides a good enough approximation. The SQL standard provides for an operator, TABLESAMPLE, to accommodate this. In Postgres, the TABLESAMPLE keyword in used in the FROM clause as follows:

```
FROM table_name TABLESAMPLE sampling_method (percent)
```

This will result in the system sampling from the table `table_name` using the sampling method `sampling_method`, until a total of `percent` of all the rows in `table_name` are returned. This sampling precedes the application of any conditions in the WHERE clause. The standard PostgreSQL distribution includes two sampling methods, BERNOULLI and SYSTEM. A similar goal can be accomplished with the `random()` function:

```
SELECT * FROM Table_Name
ORDER BY random()
LIMIT n;
```

where `n` is a positive integer. The function `random()` generates, for each row, a pseudo-random floating point value v in the range $0 \leq v < 1.0$[37] that is then used by the limit to pick up whatever rows happen to have the first (lowest) n values of v. In fact, in Postgres,

```
SELECT * FROM Dataset TABLESAMPLE SYSTEM (.1);
```

is the same as

```
SELECT * FROM Dataset WHERE random() < 0.01;
```

MySQL has not implemented TABLESAMPLE, but the approach of using the pseudo-random function (called `RAND()` in MySQL) and LIMIT will work.

- Discretization: this is a transformation that changes numerical continuous values to ordinal ones. This is a very common transformation for classification tasks. As an example, in the people dataset the attribute 'height' may be transformed from a number to a categorical attribute with values 'low,' 'medium,' and 'high,' according to certain cut points. This can be accomplished in SQL, although it requires several steps: first, since each attribute has a type, if we are going to change 'height' to categories expressed by labels, we need to create a new, string-based attribute:

```
ALTER TABLE People ADD ATTRIBUTE height-category varchar(6);
```

[37] To obtain a random integer r in the range $i \leq r < j$, one can use the expression `FLOOR(i + RAND() * (j - i))`.

and then carry out the transformation:

```
UPDATE People
SET height-category =  CASE WHEN height > 5.8 THEN 'low'
                            WHEN height > 5.2 THEN 'medium'
                            ELSE 'low' END;
```

Note that we could represent the values 'low,' 'medium,' and 'high' with numbers (say, 0, 1, and 2). However, codes are always somewhat opaque. Note also that this procedure allows us to keep the old, original values of the attribute (since this operation loses information, that is generally a good idea).

- Binarization: this is a variation of discretization, in that it takes a numerical continuous attribute or a categorical one and transforms it into a binary attribute (or several binary attributes). Reusing the previous example, 'height' could be transformed into 3 binary attributes called 'low,' 'medium,' and 'high,' each one with possible values 'yes' or 'no' only (in some systems, a BINARY or BIT type exists that can represent 'yes/no' values with 1/0). This process is also called *creating dummy variables* in statistical contexts. This transformation is sometimes needed when an algorithm explicitly calls for such variables, and it can be performed similarly to discretization. We review it in depth in the next subsection.

Sometimes, any type of attribute may benefit from some additional treatment. For instance, in highly skewed distributions we may choose to keep a specified number of the most frequent values and create a single, new value to represent the long tail of remaining values. We have seen how to do similar manipulations by combining GROUP BY and CASE to create bins for both categorical and numerical attributes.

Exercise 3.60 Assume table `Income(PersonID, yearly-income)` and assume the table is quite large and attribute `yearly-income` has a long tail of small values. Create a new table called `NewIncome` with the same schema and data as `Income` except that all tuples in the long tail are gone (pick a value `cutpoint` where the long tail starts) and a new tuple `(id, v)` is added instead, where `id` is a new, made-up id and `v` is the average of all `yearly-income` values with size in the long tail.

We have left *dimensionality reduction* and *feature creation* for last, as they are complex operations, typically not expressed in SQL (although feature creation, in simple cases, can be done without difficulty). The best way to explain these is to think of a prediction task (classification or correlation) and consider all the attributes of a dataset as divided between the predictors and the outcome (usually, only one attribute at a time is considered for outcome). One question we face is whether the predictors, considered together, contain enough information to determine the outcome. There are three possible scenarios:

- Just enough predictors: all the predictors together can determine the outcome;
- Too many predictors: some of the predictors are actually redundant or unnecessary and do not help in predicting the outcome.

- Not enough predictors: even all predictors combined together do not have enough information to predict the outcome.

In the second case, we may want to identify and get rid of the useless attributes. The reason is that, for many algorithms, more attributes mean more parameters to consider, and this implies more work to do; in the worst case scenario, these useless attributes can confuse the algorithm. There is a substantial body of research on this issue; most of it uses sophisticated techniques that are difficult or impossible to implement in SQL, like PCA (Principal Component Analysis). We will not cover them in this book.

In the third case, we may want to create new, additional attributes by combining existing ones, in the hope that the new attributes will allow us to predict the outcome by making explicit some information that was implicit in the attributes. A typical example of this is a set of data points in 2 dimensions (given by x and y) that need to be classified into one of two classes (binary classification). The data resists our attempts, so we create a new dataset with three attributes $(x^2, y^2, \sqrt{2}xy)$, which allows us to apply a simple classifier and be successful.[38] A simple transformation like this is clearly doable in SQL:

```
CREATE TABLE NewDataset AS
SELECT x*x, y*y, sqrt(2)*x*y
FROM Dataset;
```

The tricky part here, of course, is to come up with a suitable transformation. This is also an advanced topic that we do not cover.

3.4.1 Restructuring Data

Sometimes data needs to be structured for further analysis. This is especially the case when the dataset comes from several files that go into different tables or when data is not in a *tidy* format (see Sect. 2.1.4), since most Data Mining and Machine Learning tools assume that all the data is presented as a single tabular structure and that this structure is tidy.

There are three types of situations where we may want to combine data from different tables into a single one or restructure a single table. The first one involves tables with complex structures (objects with multi-valued attributes, or different kinds of related objects) that we examined in Sect. 2.2. Such tables, when connected by primary key–foreign key connections, are put together with joins.[39]

[38] For those who have seen this before: with some datasets that are not linearly separable, the transformation yields a new dataset where the points can be separated with linear regression.

[39] Sometimes, especially when data does not come from a database but from spreadsheets, files, and similar sources, the primary keys and foreign keys may not be explicit. However, in most some cases some sort of identifier attribute is used to glue the data together. We can always use joins on

The second type involves similar data that is distributed into kinds. Assume, for instance, that we have data on university rankings, but with different rankings for different specialties: we have a table

`PsychologyRank(school-name, state, type, ranking-position)`

for schools ranked according to their Psychology programs; another table

`EconomyRank(school-name, state, type, ranking-position)`

where the schools are ranked according to their Economy programs; another table

`HistoryRank(school-name, state, type, ranking-position)`

and so on. As another example, assume that we have real estate sales information from New York, but we have different datasets for each one of the 5 boroughs that are part of the city: Manhattan, Brooklyn, Queens, The Bronx, and Staten Island. On each dataset, we have a similar schema: `(address, type, date-sold, amount-sold)`. What characterizes these *distributed* datasets is that all tables have the same or very similar schema. The way to deal with such datasets is by combining them with *set operations*, explained in Sect. 5.4.

The third type of situation is a bit more complicated; it involves data that is not *tidy* (see Sect. 2.1.4). Such data has to be changed to adjust to the format that is most appropriate for analysis. Many times, this involves *pivoting*, that is, transforming a schema that includes a series of values of an attribute as distinct columns into one where such values are part of the data—or vice versa: we may need to pivot rows to columns or columns to rows.

In essence, this involves tables where the schema contains $\texttt{name}_1, \ldots, \texttt{name}_n$, with each of those being the value of an (implicit) attribute `A`. The table may contain entries like

Attribute	**name$_1$**	...	**name$_n$**
a_1	value$_{11}$	...	value$_{1n}$
a_2	value$_{21}$	...	value$_{2n}$
	...		
a_m	value$_{m1}$	...	value$_{mn}$

that we would like to be transformed into

Attribute	**A**	**Value**
a_1	name$_1$	value$_{11}$
a_1	name$_2$	value$_{12}$
	...	
a_1	name$_n$	value$_{1n}$
	...	
a_m	name$_1$	value$_{m1}$
	...	
a_m	name$_n$	value$_{mn}$

such attributes, even if they are not declared as primary keys or foreign keys. The usual difficulty in such cases is *identifying* the foreign keys.

These transformations can be written in SQL, albeit in a cumbersome manner, best explained through an example.

Example: Pivoting in SQL

Assume a table `Earthquakes(magnitude, Y2000, Y2001, Y2002, Y2003)`, where each row has a magnitude *n* and the number of earthquakes of magnitude *n* that happened in year 2000, the number of earthquakes of magnitude *n* that happened in year 2001, and so on. We would like to transform this into a table with schema `(magnitude, year, number-earthquakes)`, which is more amenable for analysis. The following query accomplishes this:

```
CREATE TABLE earthquatesTidy AS
SELECT magnitude, year,
       sum(CASE WHEN year = 2000 THEN "Y2000"
                WHEN year = 2001 THEN "Y2001"
                WHEN year = 2002 THEN "Y2002"
                WHEN year = 2003 THEN "Y2003"
                ELSE 0 END) as numberquakes
FROM earthquakes, (values(2000), (2001), (2002), (2003))
                  as temp(year)
GROUP BY magnitude, year;
```

This strange-looking query does the following:

- It generates a table `TEMP(year)` with the values being $\texttt{name}_1, \ldots, \texttt{name}_n$ (in this example, 2000, 2001, 2002, and 2003). That is, this table contains all the values of the implicit attribute `year`.
- It takes the cross-product of the dataset (`earthquakes`, in this example) and this newly generated table, thus combining each row of data with all possible values of the implicit attribute. This makes it possible to generate, for each original row of data, as many rows as values there are in the implicit attribute.
- It uses a CASE to pick, for each case of the implicit attribute, the appropriate value in the data.

Example: Pivoting

To understand this transformation well, it is a good idea to see it in action with a toy example. Assume table

Earthquakes			
Magnitude	**Y2000**	**Y2001**	**Y2002**
1.2	3	4	5
1.5	5	6	7
2	4	8	9

Table `temp` is

Year
2000
2001
2002

Their Cartesian product is

Magnitude	Y2000	Y2001	Y2002	Year
1.2	3	4	5	2000
1.5	5	6	7	2000
2	4	8	9	2000
1.2	3	4	5	2001
1.5	5	6	7	2001
2	4	8	9	2001
1.2	3	4	5	2002
1.5	5	6	7	2002
2	4	8	9	2002

The CASE runs through this; on each row, when the value of attribute `year` is 2000, it picks the value from column "Y2000" (in other columns, it picks a 0, which has the effect of skipping them), and the same for each other value of attribute `year`. Finally, values are summed across the magnitude and year.

Exercise 3.61 Recreate the example above in Postgres. That is, create a table `Earthquake(magnitude, Y2000, Y2001, Y2002, Y2003)` and insert the data shown in it. Then run the query above to see the resulting table. Tie each value on the result to the data it came from in the original table.

Exercise 3.62 Repeat the previous exercise in MySQL. Note that the syntax is going to be a bit different.

If, for some reason, we actually want to reverse this process and go from the table `EarthquakeTidy(magnitude, year, number-earthquakes)` to the original table, this can be achieved with the following query:

```
SELECT magnitude,
       sum(CASE WHEN year = 2000 THEN numberquakes
                                 ELSE 0 END) as "Y2000",
       sum(CASE WHEN year = 2001 THEN numberquakes
                                 ELSE 0 END) as "Y2001",
       sum(CASE WHEN year = 2002 THEN numberquakes
                                 ELSE 0 END) as "Y2002"
FROM earthquakesTidy
GROUP BY magnitude;
```

Note that the aggregate SUM is not really adding anything: the table
`EarthquakeTidy(magnitude, year, number-earthquakes)`
has only one row per magnitude–year combination. Since we are grouping by magnitude and picking the years apart in the CASE statement, each SUM will only use one value. But using the aggregate allows us to use a GROUP BY clause to generate a single tuple per magnitude.

This approach is clearly burdensome, and it can become unpractical when the number of values of the implicit attribute is high. In some systems, there is a built-in pivot capability, usually called the `crosstab` operator, that can achieve similar results. For example, Postgres has this built-in capability.

Crosstab in Postgres

The query

```
SELECT * FROM crosstab(
           'SELECT magnitude, year, number-earhquakes
            FROM earthquakes order by 1',
           'SELECT distinct year
            FROM earthquakes order by 1');
```

would allows us to go from a table
`Earthquakes(magnitude, year, number-earthquakes)`
to a table with schema `(magnitude, Y2000, Y2001,...)` in Postgres. Note that `crosstab` takes 2 strings as arguments, each string being an SQL query (a SELECT statement): the first string/query defines the original data table, and the second one indicates which attribute is going to be 'spread' into the schema.

Exercise 3.63 Assume a table `GRADES(student-name, exam, score)`, where attribute `exam` can be one of 'midterm' and 'final.' Produce a table with schema `(student-name, midterm, final)` that is the cross-tab of the original one.

A closely related issue is the creation of a *dummy variable* (also known as an indicator variable, design variable, one-hot encoding, Boolean indicator, binary variable, or qualitative variable). Dummy variables are "proxy" variables, numerical values made up to represent categorical or ordinal values for analysis approaches that do not handle nominal values and require numbers (for instance, regression models). Given a categorical attribute with n different values $v_1, \ldots, v_n$, n dummy variables $A_1, \ldots, A_n$ are created. When the categorical attribute has value v_i, we set $A_i = 1$ and $A_j = 0$ for $j \neq i$. Note that it is not strictly necessary to have n different attributes; we could do with one less, since the nth case can be encoded by setting all other $n - 1$ dummies to zero. For instance, a binary categorical variable (like 'male/female,' 'indoor/outdoor') can be represented by a single dummy variable with 0 for one category (say 'male' or 'indoor') and 1 for the other ('female' or 'outdoor'). However, it is customary to use n attributes (in this example, an attribute `Male` or `Indoor` and another attribute `Female` or `Outdoor`), since many data mining and machine learning algorithms expect this format.

This a similar problem to pivoting, since we want to distribute the n values of a (categorical) attribute into a schema with n attributes; the only difference is that the new values for those new n attributes are 0 or 1. This can be achieved with the same approach as above: using the SUM aggregate on the categorical variable, but passing 1 and 0 as values (since the SUM is not really aggregating anything, this will the final values too).

Example: Creating Dummy Variables

Assume a data table like

Name	Category
Jones	A
Jones	C
Smith	B
Lewis	B
Lewis	C

This is transformed by query

```
SELECT name,
        sum(CASE WHEN (category = "A", 1, 0)) AS A,
        sum(CASE WHEN (category = "B", 1, 0)) AS B,
        sum(CASE WHEN (category = "C", 1, 0)) AS C
FROM Data
GROUP BY name;
```

into the table

Name	A	B	C
Jones	1	0	1
Smith	0	1	0
Lewis	0	1	1

Again, note that since each name appears only once, there is no real SUM. The aggregate is used so we can group by the name, thereby making sure we create a unique row for each name.

Exercise 3.64 Assume as before a table `GRADES(student-name, exam, score)`, where attribute `exam` can be one of 'midterm' and 'final.' Produce a table with schema `(student-name, midterm, final)`, where attribute 'midterm' is 1 if the student had a score > 60 and 0 otherwise, and similarly for attribute 'final.'

3.5 Metadata and Implementing Workflows

The process of exploring, cleaning, and preparing data is essentially *iterative*. After some EDA, we may discover attributes that need cleaning or other processing; after this is done, we may understand the data better and carry out some further exploration and processing. As we go on, we sometimes modify the original, raw data, generating new datasets. We need to manage all these datasets and keep track of how they were created and why. The step-by-step transformation of the data is sometimes called a *workflow*, as it corresponds to a sequence of actions applied to the data or to the results of earlier actions.

Most actions can be done in two modes: destructive and non-destructive. In destructive mode, the action changes some data, so that the old version of it is lost, and a new version put in its place. In non-destructive mode, data is transformed by generating a new version, but keeping the old data. Each action in the cleaning and pre-processing state can be implemented in a destructive and a non-destructive way. Also, all actions can be classified as *reversible* and *non-reversible*. A reversible action is one that can be undone; for instance, concatenating two strings using some character c as separator can be undone (if we register the fact that c was used as the separator, and this character does not exist in the original strings). A non-reversible action, in contrast, cannot be undone: for instance, trimming whitespaces from a string cannot be undone unless we note, for each string, how many whitespaces were deleted, and where they appeared—without this, we cannot recreate the original string from the modified one. It is clear that reversible actions can be done in a destructive mode, since we can always get back the old data. However, non-reversible actions can also be reversed if implemented in a non-destructive mode: for instance, if when we trimmed the whitespaces from a string by simply generating a new string and keeping the old one, we can choose to ignore that new string and revert to using the original one.

Implementing any actions in a non-destructive manner results in more data being added to the dataset, requiring additional storage. In choosing whether an action is done in destructive or non-destructive ways, one must balance the ability to undo actions if necessary with the extra storage requirements.

In databases, a destructive action is implemented by an UPDATE statement. We modify an attribute by using a command of the form:

```
UPDATE TABLE SET ATTRIBUTE = FUNCTION(ATTRIBUTE);
```

This is called an *in-place* update because the space used by the attribute is reused for the new value; the old value is gone.

In contrast, a non-destructive action is implemented by adding the result of a change as a new attribute in the table, or creating a brand new table. The former involves a change of schema (we are adding a new attribute), so it must be done in two steps in SQL:

1. First, the schema is modified with an `ALTER TABLE` command:

```
ALTER TABLE ADD ATTRIBUTE new-attribute-name datatype;
```

   Note that a new attribute is added to each existing row in the table; since there is no value for it, a NULL is put by default in this new attribute.
2. Next, the change is made but the result is deposited in the new attribute:

```
UPDATE TABLE SET new-attribute-name = FUNCTION(ATTRIBUTE);
```

The latter (creating a brand new table) is easy:

```
CREATE TABLE NEW-TABLE AS
SELECT (all attributes of existing table except
        the one being changed),
       FUNCTION(ATTRIBUTE) AS new-attribute-name
FROM TABLE;
```

Note that this creates a copy of the whole dataset, so it should not be every time we make a change to the data, as it would result in a proliferation of tables and a duplication of all unchanged data. However, it may be a good idea to do this at certain points (after very important changes, or after a raft of related changes).

All relational systems have an additional mechanism to evolve data: a *view* is a virtual table, one that is defined through a query. That is, a view is created as follows:

```
CREATE VIEW name AS
SELECT attributes
FROM TABLE
WHERE conditions;
```

The view's schema is defined, implicitly, by the query used. The view inherits the attribute names from the table used. If new names are desired, they can be specified with

```
CREATE VIEW name(first-new-name, second-new-name,..)  AS
SELECT attributes
FROM TABLE
WHERE conditions;
```

or as

```
CREATE VIEW name AS
SELECT first-attribute as first-new-name,
       second-attribute as second-new-name,...
FROM TABLE
WHERE conditions;
```

It is also possible to apply functions to the attributes, hence creating a view that is the result of doing some cleaning/transformation to existing data.

The view does not actually have data of its own; it depends on its definition. If we want to see what is in a view, the system simply runs the query that was used in the view definition. As a result, if the tables used in the query that defined the view change, the view itself changes accordingly. Note that, since the view is not actually stored, no extra space is needed for views. On top of that, we can define views using already-defined views, so that this approach can be used for complex workflows. Thus, using views instead of tables should be considered when manipulating a dataset (see next subsection for an additional advantage of views).

3.5.1 Metadata

Some datasets may come with metadata telling us some of these characteristics; typically, the type of dataset is known in advance, and sometimes the schema is too. In this case, EDA can be used to confirm that what we have in the dataset does indeed correspond to the metadata description. When the dataset does not come with any metadata, EDA is used to create such a description. *We should always have as complete as possible an idea of what the dataset is about* before we start any serious analysis. Most analysis will require us to make some assumptions; the closer these assumptions are to the true nature of the data, the better our analysis will be (conversely, the further away our assumptions are from the data, the higher the risk of generating false or misleading results).

Whatever way we implement our workflow, it is important to keep track of what is being done. This is what metadata is for. Ideally, after acquiring the data we should have some descriptive metadata for each dataset, which we should store. If no metadata is available, we should generate our own after EDA. Next, as we go on processing the data, we should keep track of each action taken. When an existing attribute is modified, or a new one created, or a new dataset is generated, one should register the action that was taken. If this is done, it should be possible to examine each dataset in the database and have a list of all the changes that led to it; this is what we called *provenance* or *lineage* in Sect. 1.4. As explained there, keeping track of changes is fundamental for repeatability of experiments and, in general, will help us understand what is being done to the data, making the process transparent and enabling us to revisit our decisions and pursue alternatives if needed. It is also extremely helpful if the data is to be shared or published.

One very nice thing about relational databases is that *metadata can be stored in tables, just like data*. In fact, the system does this. Whenever we create a table, the system registers this fact in some special tables, sometimes called *system tables* or *catalog*. For each table, the system keeps track of its name and its schema (attribute name, type, etc.). Also, information about keys (primary, unique, and foreign) is kept, as well as information about users and their privileges (i.e. which data in the database they have access to).

Example: Metadata in Postgres

In the Postgres command line, the following provide information about the database:

- `\l` shows all schemas/databases in the server.
- `\d` shows all tables or views in the current database. `\dt` shows tables only, `\dv` shows views only. `\dp` shows all tables and views and, for each one, who has access to them.
- `\d name`: for each table or view matching `name`, it shows all columns, types, and other related information.
- `\dg` (`\du`) shows all users (called 'roles' in Postgres) of the current database.

Example: Metadata in MySQL

In the MySQL command line, the following provide information about the database:

- `SHOW DATABASES` (also `SHOW SCHEMAS`) displays the names of all databases in the server.
- `SHOW TABLES (FROM|IN) db-name` shows all tables in the database `db-name`. Also, `SHOW TABLES (FROM|IN) db-name LIKE tablename` shows all data about any table with name `tablename`.
- `SHOW CREATE TABLE tablename` displays the `CREATE TABLE` statement that generated the table `tablename`. The statement includes all schema information, as we saw in Sect. 2.4. Alternatively, command `SHOW COLUMNS FROM tablename` shows column information for all columns in table `tablename`.
- `SHOW CREATE VIEW viewname` displays the `CREATE VIEW` statement that generated the view `viewname`. As we just saw, this includes the SQL query that creates the view.

This basic metadata gives only basic information. For a given dataset, we can create a table that describes additional metadata, as given in Sect. 1.4. Thus, we could have a table with schema:

```
(attribute-name, representation, domain, provenance, accuracy,
completeness, consistency, currency, precision, certainty)
```

and each row describing one attribute of our table. The attribute `representation` should coincide with the data type used to store the data. The attribute `domain` should describe the underlying domain in terms that a person can understand; it is a good idea to make this attribute a long string or text type, so we can describe the domain in plain language. Information about good and bad values, 'typical' values and so on can also be included here. All other attributes are about the quality of the data (again, see Sect. 1.4). Note that not all attributes will require all these features; `precision`, for instance, is associated with numerical measurements. We may expect the table, then, to have some nulls on it, for reasons similar to those seen in the *Chicago employees* dataset.

Each time an action is carried out, the effect should be registered. Thus, a second table for each dataset should be added to describe actions. For each action, we want to register

- the attribute or attributes affected by the action;
- the change made (and any functions applied);
- the values of any parameters used by the functions;
- when the action was carried out;
- *why* it was carried out;
- who carried the action out.

The first three records can be captured simply by copying the SQL command used; this can be easily stored in a long string attribute. The fourth one can be very useful in case we need to determine in which order actions were carried out and/or undo some of them; using the timestamp of changes helps see what was applied to what. The fifth one is both very important and rarely registered. In fact, it may be the most important and ignored part of all metadata. Decisions taken during EDA are many times based on a partial, sometimes faulty understanding of the data; some assumptions are usually made. If we later on discover that our assumptions were incorrect, we may want to undo those actions that were guided by newly revised assumptions (and, sometimes, we may have to undo any further changes, which is where the timestamp helps). Finally, the last record may be important in scenarios where there is a need to create `audits` because access to data is restricted or because there are other reasons (legal, regulatory) to keep this information.

One nice thing about using views is that the system does keep track of their provenance automatically. When a view is created, the system stores its definition in a catalog table that is devoted exclusively to views. Hence, with views it is not necessary to register how they were created. Still, we may want to capture *why* the change was done: the system does not do that and, as we argued above, it is important information.

Chapter 4
Introduction to Data Analysis

4.1 What Is Data Analysis?

Data Analysis is the set of tools and techniques used to extract information from data. This information comes in the form of patterns, formulas, or rules that describe properties of the dataset. Because there are many types of information that can be learned from data, data analysis is a vast and complex subject with an abundant bibliography, especially about *Data Mining* and *Machine Learning* [5, 13, 14, 16]. In this chapter, we are going to concentrate on a few, simple methods that can be implemented in SQL without excessive complexity.

Most techniques described apply to tabular data (although there are also specific techniques for text and for graphs, which we also briefly describe). In these techniques, we have a set of records (records, objects, events, observations), each represented by n attributes (observations, features) $x_1, \ldots, x_n$. In the case of *supervised* learning techniques (also called *predictive* techniques), each object also has some additional feature y of interest, usually called the *dependent variable* or *response variable* in Statistics (the *label* in Machine Learning). We want to predict the value of y from the values of $x_1, \ldots, x_n$ (called the *independent, or predictor, variables* in Statistics, and simply the *features* in Machine Learning). We start with a set of records for which the values of y are known; this is called *labeled* data. We will use such values to *train* an algorithm. The supervision here refers to the fact that we supply the algorithm with examples of what we want to know through the labels, so that the algorithm can learn form these. The label can be either categorical (in which case it describes a class out of a finite set of possible classes, and the task is called *classification*) or numerical (normally a real value, in which case the task is called *regression analysis*) [4, 16].

Supervised learning is conventionally used when we have a certain analysis in mind, or a certain hypothesis that we want to examine. Having this means that we have identified a dependent or response variable among all attributes present in the dataset. Once this is done, we start by splitting data into the *training* data and the

A. Badia, *SQL for Data Science*, Data-Centric Systems and Applications,
https://doi.org/10.1007/978-3-030-57592-2_4

test data. The training data, with labels included, is supplied to the algorithm. Once the algorithm is trained, it is run on the test data (with labels withheld). We thus can compare the answers that the algorithm provides on the test data with the labels for such data to determine how well the algorithm is performing. For this to work well, the data should be split randomly, to avoid providing the algorithm with training data that is biased in some way. There are some smart ways to split data, but here we will use a simple method that relies on a random number generator.

Sometimes we do not have any particular attribute to serve as the dependent variable; this is the *unsupervised* learning case (also called *knowledge discovery*). In this case, we want to explore the data and find patterns on it, without much in the way of assumptions. Unsupervised learning can sometimes be used as part of the pre-processing of data, since it can help us learn more about the dataset and its lack of assumptions fits this stage of analysis well. For instance, a very common task is to discover whether the data records we have can be divided into groups based on their similarities to each other. This is called *clustering*, and the groups of similar objects we find are called *clusters*. Other types of unsupervised learning include finding *association rules* and discovering *latent factors* (also called *dimensionality reduction*).

There are also other approaches, like *semi-supervised* learning. There are methods that combine more than one tool, called *ensemble* models.

Here we are going to cover only some basic ideas that can be implemented in SQL with relatively limited effort. Knowing these ideas is quite useful: they can serve as an introduction to more complex approaches. From a practical perspective, it is important to start analysis with simple tools and only use more complex methods once the data (and the problem) are well understood. For many real-life problems, a simple method may provide approximate but useful results.

4.2 Supervised Approaches

In supervised approaches, we have labeled data that we need to split into training data and test data. One good way to do this is to choose randomly; as we have seen (see Sect. 3.4), most systems have a way to generate random numbers that can be used for this. Recall that in Postgres, for instance, the function `random()` generates a random number between 0 and 1 each time it is called (so it generates multiple random numbers, if called several times in the same query), and that MySQL has the exact same function. It is common to devote most data (between 75% and 90%) to training and a small part (between 25% and 10%) to testing. A general procedure to split dataset into training and test would be written (in Postgres or MySQL) as follows:

```
CREATE TABLE new-data
as SELECT *, random() as split
FROM Data;
```

```
CREATE TABLE training-data
as SELECT *
FROM new-data
WHERE split >= .1;

CREATE TABLE testing-data
as SELECT *
FROM new-data
WHERE split < .1;
```

Remember that `random()` simply generates a random value each time it is called; and in the first query above, it is called once for each row in the `Data` table. Thus, the table `new-data` is simply rows of data with a random number between 0 and 1 attached to each row. By changing the constant in the definition of table `training-data` (and adjusting the constant in the definition of `testing-data`), we can split the data as needed.

4.2.1 Classification: Naive Bayes

In classification, we assume that each record has n predictive attributes $(A_1, \ldots, A_n)$ and belongs to one of several classes $C_1, \ldots, C_m$. In the training data, each record has an attribute `Class` giving the class C_i of that record, so the schema of the training data is (A_1,...,A_n,`Class`). We create the training data as shown above and train our algorithm on it. We then take out attribute `Class` from the test data and run our trained algorithm in this data to see which class it assigns to each testing record. We then compare the predicted class to the real, withheld class to see how well our algorithm did.

Classification is one of the most studied problems in Data Science, and there are many algorithms to attack it. One of the simplest is *Naive Bayes*. In spite of its simplicity, it can be surprisingly effective, and in simple situations it can be written in SQL.

Assume we have a table `training-data`(A_1,...,A_n,`Class`), where each A_i is an attribute (feature) that we are going to use for classification. The idea of Naive Bayes is this:

1. For each class C_i, the *a priori probability* of the class, $P(C_i)$, is computed as the percentage of records in the dataset that belong to this class. We can easily compute this for each class; we compute also the raw count of records for each class because it will be useful in the next step:

```
CREATE TABLE classPriors AS
SELECT class, sum(1.0 / total) as classProb,
            count(*) as rawclass
FROM training-data,
     (SELECT count(*) AS total FROM training-data) AS T
GROUP BY class;
```

This yields a table with schema `(class, classProb)`, with one row for each value of `Class`; row (C_i, p) means that class C_i has probability p (p will always be a real number between 0 and 1).

2. For each predictor attribute A_i, class C_j, we compute the conditional probability $P(C_j|A_i = a_i)$ that, if attribute A_i has value a_i, the record is of class C_j. This is done for each value of A_i, so again it is a simple grouping:

```
CREATE TABLE AiPriors
SELECT Ai, class, sum (1.0 / rawclass) AS classProb
FROM training-data, classPriors
WHERE training-data.class = Priors.class
GROUP by Ai, class;
```

This yields a table with schema (A_i,`class`,`classProb`) with tuples (a_i, C_j, p) reflecting that $P(C_j|A_i = a_i) = p$ (as before, $0 \leq p \leq 1$). Note that we divide by the number of records in the class because what matters to us is what is the influence of the fact that attribute A_i has value a_i in the fact of the record being of class C_i; therefore, we need to take all and only the records of class C_i as reference.

A similar table is created for each predictor attribute, so we get a table for A_1, a table for $A_2, \ldots$, and a table for A_n.

3. Given all these prior probabilities, we can now apply our predictor as follows: to predict the probability that a record $r = (a_1, \ldots, a_n)$ in the testing set belongs to class C_i, we use

$$P(C_i|r) = P(C_i|a_1, \ldots, a_n) = \Pi_j P(C_i|A_j = a_j)P(c_i).$$

Note that this assumes that the probability of each single attribute denoting a class is independent of all other attributes, a strong assumption that does not always hold (hence the 'naive' in 'naive Bayes'). In spite of this, the approach works unexpectedly well in many situations, including some with dependencies among the predictors.

This calculation is carried out for each class, that is, in each training record r we get a probability $P(C_i|r)$ for each of $C_1, \ldots, C_m$. We then choose the class with the highest probability as the class of r. The process is repeated for each record in the testing data:

```
CREATE TABLE results as
SELECT test-data.*, classPrior.class,
    classPrior.prior * A1prior.prob * A2prior.prob ...
    as ClassProb
FROM classPriors, A1Prior, ..., AnPrior, test-data
WHERE test-data.A1 = A1Prior.value
      and classPrior.class = A1Prior.class and
      test-data.A2 = A1Prior.value
      and classPrior.class = A2Prior.class and
      ...
GROUP BY test-data.*, classPrior.class;
```

We have used '*' here as a shortcut; it is necessary to enter all attributes of `test-data` to make the query legal. That is, the schema here is `(`A_1`,...,`A_n`,Class, ClassProb)`, with each test record $r = (a_1, \ldots, a_n)$ generating m records of the form $(a_1, \ldots, a_n, C_i, p)$ (one for each C_i), meaning that $P(C_i|r) = p$. This step is done for each record in the testing set and each class; now we chose our final results:

```
SELECT test-data.*, classPrior.class
FROM results R1
WHERE ClassProb = (SELECT max(ClassProb)
            FROM results R2
            WHERE R2.test-data.* = R1.test-data.*)
```

Again, the '*' stands for all predictor attributes $A_1,\ldots,A_n$.

Once this is done, the results are compared with the real class of those records; the percentage of correct predictions is a good indication of how good our Naive Bayes classifier is.

Example: Naive Bayes in SQL

Assume a dataset where we have information about several patients and their eyesight issues.[1] In particular, a table `RawData` has schema `(id, age, prescription, astigmatic, tears, lens)`. We are going to predict attribute 'lens' from attributes 'age,' 'prescription,' 'astigmatic,' and 'tears.'

```
%split data into training and testing, as indicated
CREATE TABLE FullData AS
SELECT *, rand() as split
FROM RawData;

CREATE TABLE TrainData AS
SELECT id, age, prescription, astigmatic, tears, lens
FROM FullData
WHERE split >= .01;

CREATE TABLE TestData AS
SELECT id, age, prescription, astigmatic, tears, lens
FROM FullData
WHERE split >= .01;

%calculate class priors
CREATE TABLE Priors AS
SELECT lens, count(*) as rawclass, sum(1.0 /t.total) as prior
FROM TrainData,
    (SELECT count(*) as total FROM TrainData) AS t
GROUP BY lens;

%calculate conditional probabilities per attribute and class.
```

[1] This example is adapted from the one at https://sqldatamine.blogspot.com/.

```
CREATE TABLE AgeCondProbs AS
SELECT age as value, TrainData.lens,
       sum(1.0/rawclass) as condprobs
FROM TrainData, Priors
WHERE TrainData.lens = Priors.lens
GROUP BY age, lens;

CREATE TABLE PrescriptionCondProbs AS
SELECT prescription as value, TrainData.lens,
       sum(1.0 / rawclass) as condprobs
FROM TrainData, Priors
WHERE TrainData.lens = Priors.lens
GROUP BY prescription, lens;

CREATE TABLE AstigCondProbs AS
SELECT astigmatic as value, TrainData.lens,
       sum(1.0/ rawclass) as condprobs
FROM TrainData, Priors
WHERE TrainData.lens = Priors.lens
GROUP BY astigmatic, lens

CREATE TABLE TearsCondProbs AS
SELECT tears as value, TrainData.lens,
       sum(1.0/rawclass) as condprobs
FROM TrainData, Priors
WHERE TrainData.lens = Priors.lens
GROUP BY tears, lens;

%using probabilities for class and attributes,
%compute the class of each record in training data
CREATE TABLE Results as
SELECT id, Priors.lens,
       (A.condprobs * B.condprobs * C.condprobs * D.condprobs
       * Priors.prior)  as classProb
FROM TrainData, Priors
     AgeCondProbs A, PrescriptionCondProbs B,
     AstigCondProbs C, TearsCondProbs D
WHERE TrainData.age = A.value and
      TrainData.prescription = B.value and
      TrainData.astigmatic = C.value and
      TrainData.tears = D.value and
      Priors.lens = A.lens and Priors.lens = B.lens and
      Priors.lens = C.lens and Priors.lens = D.lens
GROUP BY id, Prior.lens;

%we chose the class here as the one with highest probability
CREATE TABLE Eval as
SELECT Results.id, Results.lens
FROM  Results R1
WHERE classProb = (SELECT max(classProb)
                   FROM Result.R2
                   WHERE R1.id = R2.id)

%evaluation of results: compare to real class (ground truth)
```

```
SELECT sum(case when Eval.lens = TestData.lens
          then 1 else 0 end)/count(*)
FROM TestData, Eval
WHERE TestData.id = Eval.id;
```

Exercise 4.1 One of the most common exercises in all of Machine Learning is to apply some simple algorithms (like Naive Bayes) to the *iris dataset*, a very famous (and simple) dataset available in many places on the Internet.[2] Find a copy of the dataset, load it into Postgres or MySQL, and implement the Naive Bayes algorithm over it (what features are the predictive ones and which one is the class to be predicted will be obvious once you learn about the dataset). Pick 90% of the data randomly for training and 10% for validation.

While the implementation above is straightforward, there is a practical issue to consider: when the number of predictive attributes is high, this requires quite a bit of tables (and queries). Note that the priors of the classes can be computed with a single query, regardless of the number of classes; this is due to the fact that what we want (the particular classes) are in the data, while the (predictive) attributes are part of the schema. Therefore, in some approaches the calculated table is 'pivoted' (see Sect. 3.4.1) so that all per-attribute probabilities can be calculated in a single query (albeit a pretty long one) and put in a single table. However, this means that all probabilities will fall under one attribute. In this case, since in SQL there is no multiplication aggregate, we cannot write

```
SELECT attribute, mult(probs)
...
GROUP BY attribute;
```

We instead use the trick described earlier (see Sect. 3.1): we add logs instead of multiplying probabilities and then take the obtained value as an exponential. That is, instead of multiplying numbers $r_1 \times \ldots \times r_n$, we add $log(r_1) + \ldots + log(r_n)$ (which uses the aggregate `sum()`) and then compute e^r (for r the result of the sum), which in most systems is done with function `exp()`. Thus, we write

```
SELECT attribute, exp(sum(log(probs)))
...
GROUP BY attribute;
```

instead of the above.

As stated earlier, we must realize that the calculated value is an approximation due to numerical representation limits; to minimize loss of accuracy, it is important that the value be represented as a real number with the biggest precision available in the system. Even then, very small numbers (as probabilities tend to be, since they

[2] See https://en.wikipedia.org/wiki/Iris_flower_data_set for a description and pointers to several repositories of the data.

are normalized to between 0 and 1) may sometimes render inaccurate results. This is especially the case with large datasets where the class data is sparse (i.e. each class appears infrequently in the data); this makes the class priors to be very small numbers.

Exercise 4.2 Compute the Naive Bayes of the example, but this time create a single table `AttributeCondProbs(attribute, value, lens, condprob)`, where `attribute` is one of the attribute names ("age," "prescription," "astigmatic," "tears"), `value` is the value of the given attribute, `lens` is one of the classes, and `condprob` is the conditional probability that when attribute `attribute` takes value `value` in a record, and the record is of class `lens`. Hint: you can union the individual tables computed above. Another hint: when you do this, you can apply the trick of adding logs to compute the final conditional probability.

We mention an additional feature of Naive Bayes. When the model has many features, it could be that the estimated probability of one (or several) of the features for some class is zero (i.e. there is no record where attribute A_i has a certain value a_i for some class C_j). The problem with this is that it yields a conditional probability of zero; as we have seen above, we are going to multiply probabilities, so if one of them is zero, that will bring the whole product to zero. For this reason, it is common to use a technique called *additive smoothing* or *Laplace smoothing* when calculating probabilities by counting. Without going into the technical details, the basic idea is to add a very small value (usually 1) to each count, while adding a corrective factor (usually n, for n the total number of possible values) to the denominator of the fractions that compute probabilities (otherwise, the probabilities for the values of the class may not add up to 1). Thus, for attribute A_i, we would compute $P(C_j|A_i = a_i)$ as follows:

```
CREATE TABLE AiPriors
SELECT Ai, class,
       sum (1.0 / (rawclass + norm)) + 1 AS classProb
FROM training-data, classPriors,
     (SELECT count(distinct Ai) as norm FROM training-data)
      as T
WHERE training-data.class = Priors.class
GROUP by Ai, class;
```

Recall that the fact that we do not have a multiplicative aggregate in SQL forces us to use sums of logs of probabilities. While the sum has no problem with zeros, the log function is undefined for a value of zero, so this issue does not go away in SQL; thus, this technique may have to be applied in certain problems anyways.

Exercise 4.3 Redo the exercise on the Iris dataset, but this time apply Laplace smoothing. Check the difference between your result here and without the smoothing. Did the algorithm perform better with or without smoothing?

4.2.2 *Linear Regression*

Regression is similar to classification, in that we have, in the data, independent variables or predictors, and we want to predict values of a dependent variable or response. However, in regression the response is a numerical value. For example, assessing the risk of a borrower for a loan based on attributes like age, occupation, expenses, credit history, etc., is a typical regression task, since we give each borrower a numerical *score* that represents how high/low of a risk it is to loan money to that individual. The predictors themselves can be numerical or categorical; whenever a predictor is categorical, it is transformed into a numerical one by creating a *dummy variable*, as seen in Sect. 3.3.1.1.

The simplest type of regression is *linear* regression, where one attribute (independent variable) A is related to attribute (dependent variable) B with a linear relation

$$B = f(A) = \alpha_0 + \alpha_1 A,$$

where α_1, α_2 are the parameters we need to estimate from the data (traditionally, α_2 is called the *intercept* and α_1 is called the *slope*). With n independent variables, we have a linear relation

$$B = f(A_1, \ldots, A_n) = \alpha_0 + \alpha_1 A_1 + \ldots + \alpha_n A_n$$

and try to estimate $\alpha_0, \ldots, \alpha_n$. We can start with a guess and then work to minimize the error, that is, the difference between predicted and actual value. Since this is supervised learning, on each record r we know $r.B$ (the value of B at r) , which we compare with $f(r.A)$ (the value derived from A at r): the error in the record r is usually taken to be $r.B - f(r.A)$, since we are dealing with numerical values. We want to minimize the total error for the whole dataset; usually, this is expressed as the *sum of square differences*. For the one variable case, this is given by

$$SSE = \Sigma_i(f(r_i.A) - r_i.B)^2 = \Sigma_i((\alpha_0 + \alpha_1 r_i.A) - r_i.B)^2,$$

where we sum over all records in the dataset. We simply choose the values of α_0, α_1 that minimize this. Because this is a very simple expression, we can determine what are the optimal values for these parameters:

$$\alpha_1 = \frac{\Sigma_i(r_i.A - \bar{A})(r_i.B - \bar{B})}{\Sigma_i(r_i.A - \bar{A})^2}, \tag{4.1}$$

$$\alpha_0 = \bar{B} - \alpha_1 \bar{A}, \tag{4.2}$$

where $\bar{A}$ is the mean of the A values and $\bar{B}$ is the mean of the B values. There are equivalent formulas for the slope:

$$\alpha_1 = \frac{(\Sigma_i r_i.A\ r_i.B) - (N\bar{A}\bar{B})}{(\Sigma_i r_i.A^2) - (N\bar{A}^2)} \tag{4.3}$$

or

$$\alpha_1 = \frac{N\Sigma_i(r_i.A\ r_i.B) - (\Sigma_i r_i.A)(\Sigma_i r_i.B)}{N\Sigma_i(r_i.A^2) - (\Sigma_i r_i.A)^2} \tag{4.4}$$

where N is the number of data records in the dataset. There are also equivalent formulas for the intercept:

$$\alpha_0 = \frac{1}{N}(\Sigma_i r_i.B - \alpha_1 \Sigma_i r_i.A). \tag{4.5}$$

Looking at formula 4.1 for α_1, it is clear that the top is the covariance of A and B, and the bottom is the variance of A, that is, α_1 can also be expressed as

$$\alpha_1 = \frac{Cov(A, B)}{Var(A)} = Corr(A, B)\frac{stdev(B)}{stdev(A)}. \tag{4.6}$$

All these formulas can be expressed quite easily in SQL: in a system like Postgres, where Covariance and Variance come as standard functions, formula 4.6 is also the easiest. But even in systems without these functions, writing the formulas in SQL is a matter of putting together their pieces:

- $\Sigma_i(r_i.A - \bar{A})$ is simply the variance, the sum of the differences between values of A and their mean.
- $\Sigma_i(r_i.A - \bar{A})(r_i.B - \bar{B})$ is simply the product of the variances, or the sum of the product of the differences between values of A and their mean and values of B and their mean.
- $(\Sigma_i r_i.A)$ simply requires us to sum the values of A.
- $(\Sigma_i r_i.A\ r_i.B)$ requires us to multiply, on each row, the value of A times the value of B, and sum the results.
- $(\Sigma_i r_i.A^2)$ requires that we square the value of A and add the results—note that this is different from $(\Sigma_i r_i.A)^2$, where we first add the values of A and then square the result.

For simplicity, we can do these calculations in the FROM clause and compute α_1 before α_0 since the value of α_0 can be derived from that of α_1. Our dataset consists of records that contain attributes A and B. For Definitions 4.3 and 4.5, we get

```
SELECT ((SB * SAA) - (SA * SAB)) /
         ((N * (SAA)) - (SA * SA)) AS intercept,
       ((N * SAB) - (SA * SB))  /
```

```
    ((N * SAA) - (SA * SA)) AS slope
FROM (SELECT sum(A) AS SA,
             sum(B) AS SB,
             sum(A * A) AS SAA,
             sum(A * B) AS SAB,
             count(*) AS N
      FROM Data);
```

Example: Simple Linear Regression

Assume a real estate dataset with schema:

`(Id, Address, size, price, num-beds, num-baths)`

with an Id and an address for each property, together with the size in square feet and the price paid for it the last time it was sold, as well as the number of bedrooms and the number of bathrooms in the house.

We believe that the price of a house is a direct result of its size. Then we could calculate a simple regression with the size as the predictor and the price as the dependent variable.

```
SELECT
  ((SumPrice * sumPriceSq) - (SumSize * sumSizePrice)) /
    ((N * (sumPriceSq)) - (SumSize * SumSize)) AS intercept,
  ((N * sumSizePrice) - (SumSize * SumPrice))  /
    ((N * sumPriceSq) - (SumSize * SumSize)) AS slope
FROM (SELECT sum(size) AS SumSize,
             sum(price) AS SumPrice,
             sum(size * size) AS sumPriceSq,
             sum(size * price) AS sumSizePrice,
             sum(price * price) AS SumPriceSq,
             count(*) AS N
      FROM Data);
```

Exercise 4.4 Modify the previous example to compute α_1 and α_0 using Definitions 4.1 and 4.2. Apply it to the New York real estate sales dataset using `GROSS SQUARE FEET` and `SALE PRICE`.

Exercise 4.5 Modify the previous example to compute α_1 and α_0 using Definitions 4.4 and 4.2. Apply it to the New York real estate sales dataset using `GROSS SQUARE FEET` and `SALE PRICE`.

What if we want to use more than one independent variable? Unfortunately, the formula for the general case is quite complex. We show here the 2-variable case, for which it is still possible to give a somewhat reasonable SQL query. In this case, we are looking at an equation:

$$B = \alpha_0 + \alpha_1 A_1 + \alpha_2 A_2,$$

where A_1 and A_2 are the predicting attributes (independent variables), B the predicted attribute (dependent variable), and $\alpha_0, \alpha_1, \alpha_2$ the parameters to be learned from the data. To calculate these parameters, we proceed in three steps:

1. Compute $SUM(A_1)$, $SUM(A_2)$, $SUM(A_1^2)$, $SUM(A_2^2)$, $SUM(B)$, $SUM(A_1 \cdot B)$, $SUM(A_2 \cdot B)$, and $SUM(A_1 \cdot A_2)$.
2. Using the previous results, compute
   - $SA_1^2 = SUM(A_1^2) - \frac{SUM(A_1)^2}{n}$;
   - $SA_2^2 = SUM(A_2^2) - \frac{SUM(A_2)^2}{n}$;
   - $SA_1B = SUM(A_1 \cdot B) - \frac{SUM(A_1)SUM(B)}{n}$;
   - $SA_2B = SUM(A_2 \cdot B) - \frac{SUM(A_2)SUM(B)}{n}$;
   - $SA_1A_2 = SUM(A_1 \cdot A_2) - \frac{SUM(A_1)SUM(A_2)}{n}$;
3. Finally, compute

$$\alpha_1 = \frac{(SA_2^2)(SA_1B) - (SA_1A_2 \cdot SA_2B)}{(SA_1^2)(SA_2^2) - (SA_1A_2)^2},$$

$$\alpha_2 = \frac{(SA_1^2)(SA_2B) - (SA_1A_2 \cdot SA_2B)}{(SA_1^2)(SA_2^2) - (SA_1A_2)^2},$$

$$\alpha_0 = SUM(B - \alpha_1A_1 - \alpha_2A_2).$$

Note that, in spite of the complexity of the formula, many factors are reused.

As before, this can be put into an SQL query by using subqueries in FROM to carry out computations in a step-by-step manner.

Exercise 4.6 Write an SQL query on real estate dataset to apply linear regression to number of bedrooms and number of bathrooms in a house to predict its price.

Linear regression is a well-known, simple technique, so it is often tried first on many datasets. However, it has some severe limitations of which we should be aware. First and foremost, it will always 'work,' in the sense that it will always yield an answer (the one that minimizes the error, as defined above). However, even this answer may not be very good; that is, the error it produces may still be quite large. This is because there are two strong assumptions built into linear regression: first that there is a *linear* relation between independent and dependent variables, not a more complex one.[3] Second, it assumes that the errors (and there will always be errors; perfect fits are extremely unlikely with real data[4]) behave very nicely (technically speaking, errors must be independent and identically distributed with

[3]This is a very strong assumption: see https://en.wikipedia.org/wiki/Anscombe's_quartet.

[4]In fact, perfect fits are usually a cause for suspicion and one of the reasons some cheaters have been caught.

a normal distribution). Regression will not warn us that the relationship we are looking for is not there or is not linear. How do we know that our regression result is any good?

Recall that we minimized the sum of square errors, SSEs, in our calculation. Another measure of error is the *regression sum of squares*:

$$SSR = \Sigma_i (f(r_i.A) - \bar{B})^2.$$

This measure tells us how far our estimates are from the average of the real data value (the variance of the estimates). But to make sense of this, we need to know the variance of the real values:

$$SSTO = \Sigma_i (r_i.B - \bar{B})^2.$$

The key is to note that $SSTO = SSR + SSE$. Because we can think of SSE as the variance of the data, we can test to see how much of it we account for with the regression; this value is usually called *R-square* (R^2) and defined as

$$R^2 = \frac{SSR}{SSTO} = 1 - \frac{SSE}{SSTO}.$$

The value of R^2 is between 0 and 1 (although they are many times expressed as a percentage). A high value means that most of the variance on the predicted attribute B is accounted for by the variance in the predictor attribute A. In the context of linear regression, it means that the slope and intercept obtained fit the data quite well.

An important note: the Pearson *correlation coefficient* r that we saw in Sect. 3.2.2 turns out to be the square root of R-square, $r = \sqrt{R^2}$.

Exercise 4.7 Compute the R-squared value for your results predicting the price of a house from its size on the New York real estate dataset (any version).

Another issue with linear regression is that it requires that we normalize all our data before trying it; if an attribute is of much larger magnitude than others, it will dominate the calculations, creating the biggest differences—hence, it will be minimized even at the expense of other factors. For instance, assume we are again trying to predict the price at which a house will sell and that we decide to use, this time, the size, number of bedrooms, and number of bathrooms. It is clear that the number of bedrooms and bathrooms are going to be very small numbers (from 1 to 5 or so), while the size will be a number in the hundreds (if expressed in square meters) or the thousands (if expressed in square feet), perhaps even more. If we try to use all these attributes to predict the price (and it would seem reasonable to do so), we need to normalize the size attribute or it will dominate the calculations (this is sometimes called *feature scaling* in Machine Learning).

Finally, linear regression is very sensitive to outliers. An extreme value creates a very large error and, as in the case of large magnitude attributes, the approach will try to minimize this error even at the expense of other values or other attributes. Therefore, it is very important to check the data for outliers before applying linear regression.

4.2.3 Logistic Regression

In spite of its name, logistic regression is actually a classification algorithm. However, we describe it after linear regression because it uses linear regression at its core.

The idea of logistic regression is to estimate a linear regression but to interpret it as the *odds* of the probability that the response variable is in a certain class. In the simplest case, assume our response variable is binary, so it can belong to one of two classes, 0 or 1 (for example, whether a person is a good or bad risk for loans, or whether a student will pass or fail an exam, or a patient will survive a certain procedure, etc.). If p is the probability of being of class 1 (so that $1 - p$ is the probability of class 0), the odds of being of class 1 is $\frac{p}{1-p}$. Linear regression with predictors $A_1, \ldots, A_n$ aims to approximate $Pr(1|A_1, \ldots, A_n)$. Instead, logistic calculates the log of the odds:

$$log \frac{Pr(1|A_1, \ldots, A_n)}{(1 - Pr(1|A_1, \ldots, A_n))}.$$

It does this by assuming that these odds are the product of a linear regression. In simple form,

$$log \frac{p}{1-p} = \alpha_0 + \alpha_1 A_1 + \alpha_2 A_2 + \ldots + \alpha_n A_n,$$

where, as before, the $\alpha_0, \ldots, \alpha_n$ are the parameters we want to estimate for the predictor attributes $A_1, \ldots, A_n$. Note that if we can find out the values of $\alpha_0, \ldots, \alpha_n$, then we can find out p simply by reversing the above, which comes out (after a bit of algebra) as

$$p = Pr(1|A_1, \ldots, A_n) = \frac{1}{1 + exp(-\alpha_0 - \alpha_1 A_1 - \ldots - \alpha_n A_n)}.$$

For the case of one variable,

$$p = Pr(1|A) = \frac{1}{1 + exp(-\alpha_0 - \alpha_1 A)}.$$

Thus, the idea is to compute the parameters α_0 (intercept) and α_1 (slope) by using linear regression and then use them in the formula above. This formula is also called the *sigmoid* function, and it gives values between 0 and 1. Usually, this approach is used in binary classification (i.e. deciding between two classes), with values between 0 and 0.5 are interpreted as denoting one of the classes, and values between 0.5 and 1 denoting the other class. Clearly, this can be added to our previous computation of b (intercept) and a (slope):

```
SELECT 1.0 / (1.0 + exp(-intercept * -slope * A)
FROM ...
```

where A is the attribute we are using to predict our result, and `slope` and `intercept` are the values computed using linear regression in the previous subsection.

Exercise 4.8 Assume that all we want to know about the real estate dataset is whether a house will sell for a high price (defined as more than $300,000) or a low price (defined as equal to or less than $300,000). Apply logistic regression to this problem using only size as predictor. Hint: first add a variable `classprice` to the dataset with values 1 (for high) and 0 (for low) depending on whether the house's price is above or below our threshold of $300,000 (we have seen how to do this in Sect. 3.3.1.1). Then pick some data randomly as training and apply linear regression to get a slope and intercept, which are then finally fed to the sigmoid function, as shown above.

4.3 Unsupervised Approaches

In unsupervised approaches, we are given a dataset without labels. The goal is to discover pattern or structure in the data without any guidance. For instance, instead of classifying records (called *data points* in this context) into a prefixed list of classes, we check to see if a set of potential classes emerges from the data itself. Classes obtained from the analysis of data are called *clusters*, and the task of finding such clusters is, naturally, called *clustering*. Note that this can be harder than classification, since we do not know anything a priori: the number of classes, or their nature. In general, unsupervised approaches are considered weaker than supervised ones, as it is to be expected since unsupervised approaches have less information to work with. However, since they do not require labeled data, unsupervised techniques are very important, as they can be applied to any dataset. They can also be very helpful in understanding the data, and some authors classify techniques like clustering and dimensionality reduction with EDA.

4.3.1 Distances and Clustering

Many approaches rely on defining how similar (or dissimilar) two records are and comparing all records in a dataset. Then, records that are similar enough to each

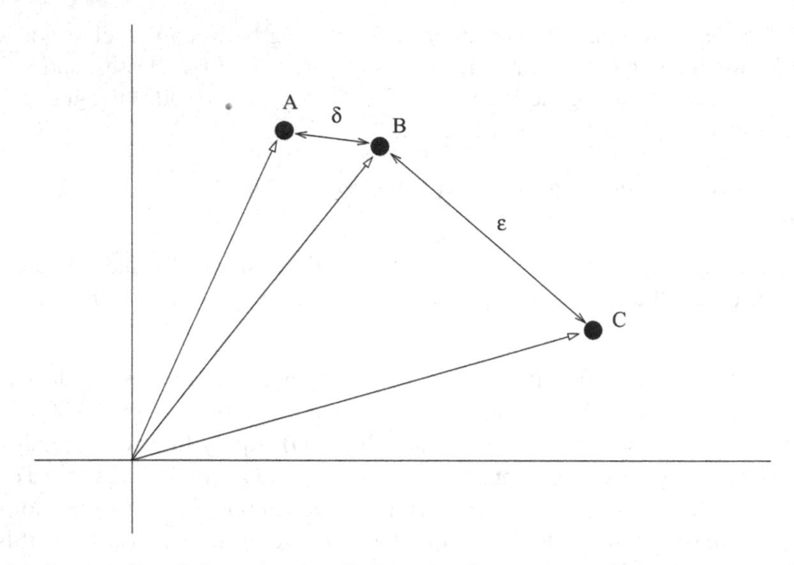

Fig. 4.1 Items A, B, and C as points in a (2-dimensional) space (or, equivalently, vectors); δ is the Euclidean distance between A and B and ϵ the Euclidean distance between B and C

other are grouped together to form a cluster. The intuitive idea is that similar records end in the same cluster, and dissimilar records end up in different clusters.

To implement this, we rely on the idea of *distance*, a function that assigns a number to each pair of records/data points. This function abstracts from the familiar idea of distance in geometry, with records seen as points in a space (hence the name; see Fig. 4.1). Intuitively, records/points that are 'close' to each other (the distance between them is small) are similar, while those that are far away according to the distance are dissimilar. Formally, a distance $D(x, y)$ is any function that applies to two data points x and y and fulfills the following:

- $D(x, y) \geq 0$ (the result is always positive), with $D(x, y) = 0$ only if $x = y$;
- $D(x, y) = D(y, x)$ (the function is symmetric; that is, order of arguments is not important);
- $D(x, y) \leq D(x, z) + D(z, y)$ (the 'triangle inequality.' Intuitively, the shortest distance from x to y is a straight line).

Distances between records/data points are usually obtained by combining distances between attributes. The typical distance for numerical attributes is the difference: the distance between numbers n_1 and n_2 is $\mid n_1 - n_2 \mid$. For categorical attributes, it is difficult to establish a meaningful distance. In some contexts, string similarity (as seen in Sect. 3.3.1.2) works well, but in many others the only distance that can be used with categorical attributes is the *trivial* distance:

$$D(s_1, s_2) = \begin{cases} 0 \text{ if } s_1 = s_2 \\ 1 \text{ otherwise} \end{cases}.$$

Distances are mostly used in the context of records with all attributes being numerical.

To combine individual distances into a distance between the two records, a very common approach is the Minkowski distance, defined as follows: for two data points x and y defined over attributes/features $A_1, \ldots, A_n$ (so that each data point is a record of values $(a_1, \ldots, a_n)$), their Minkowski distance is

$$d(x, y) = \sqrt[k]{\Sigma_{i=1}^{n} (x.A_i - y.A_i)^k},$$

where we use $x.A_i$ to denote the value of record/data point x for A_i, and likewise for $y.A_i$. Here, the number $k > 0$ is a parameter; in actuality, the Minkowski distance is a family of distance functions, one for each value of k. Some very well-known examples are the *Manhattan* distance, which uses $k = 1$:

$$d(x, y) = \Sigma_{i=1}^{n} (x.A_i - y.A_i)$$

and the *Euclidean* distance, which uses $k = 2$ (again, see Fig. 4.1):

$$d(x, y) = \sqrt{\Sigma_{i=1}^{n} (x.A_i - y.A_i)^2}.$$

These distances are quite useful, but they only work well under certain circumstances. First, they require that all the attributes have been scaled; otherwise, if one of them is much larger than others, it will dominate the distance. Recall the real estate example: a dataset with information about houses, including size, number of bedrooms, and number of bathrooms. If we are trying to determine how 'similar' two houses are, the Euclidean distance is not a bad idea, but before using it we need to normalize our data. Otherwise, any distance that combines all these attributes and does not pre-process them to remove this scale will be almost exclusively based on size and will be unable to distinguish between two houses of similar size but one with only one bathroom and another one with two or three.

The second problem with these distances is that they work better when the attributes are independent (or at least, uncorrelated). This problem is addressed by another famous distance, the *Mahalanobis* distance. This distance tries to normalize the differences between the values in each attribute by factoring in the covariances of the attributes. The idea is that if the covariance is high in absolute value, the attributes are highly correlated and any similarity should be discounted, as is to be expected. In contrast, if the covariance value is low, the attributes are closed to independent, so their distance should be considered important. To fully compute the Mahalanobis distance requires the covariance matrix of the dataset (a matrix whose element in the (i, j) position is the covariance between the i-th and j-th attributes of the dataset). This is hard (but not impossible) to compute in SQL; but then Mahalanobis requires the inverse of this matrix, which is not computable in SQL without using functions and some advanced tricks (and even then it is a real pain). But note that the covariance of any element with itself is simply its variance;

therefore, the diagonal of the covariance matrix is simply the variance of each attribute. Thus, one way to approximate the idea is to divide the distance between two values of an attribute by the standard deviation of that attribute (which is the square root of the variance, and for which a function exists in most SQL systems). Thus, we have a 'poor person' Mahalanobis distance:

$$d(x, y) = \Sigma_{i=1}^{n} \sqrt{\frac{(x.A_i - y.A_i)^2}{stddev(A_i)}}.$$

This is also called the *standardized Euclidean distance*, since it is equivalent to a Euclidean distance that uses standardized (z-score) values.

Another famous distance function is *cosine similarity*, inspired by linear algebra. Consider again each data point as literally a point in a space with n dimensions (one per attribute)—or, equivalently, as a vector (a vector is an object with a magnitude and a direction; these are usually depicted as arrows, with the length being the magnitude and arrowhead giving the direction): think of an arrow from the 'origin' of the space (the point (0,0,...)) to the point itself. Two points represent two vectors with a common origin; thus, they create an angle between them. The idea is that two similar points are very close to each other, so the angle between their vectors is very small (see Fig. 4.2). If they are pointing in exactly the same direction, the angle is 0° (so the cosine of the angle is 1); if they are at 90 degrees from each other, the angle is 90° (so the cosine is 0); if they are in the same 'line' but opposite directions, the

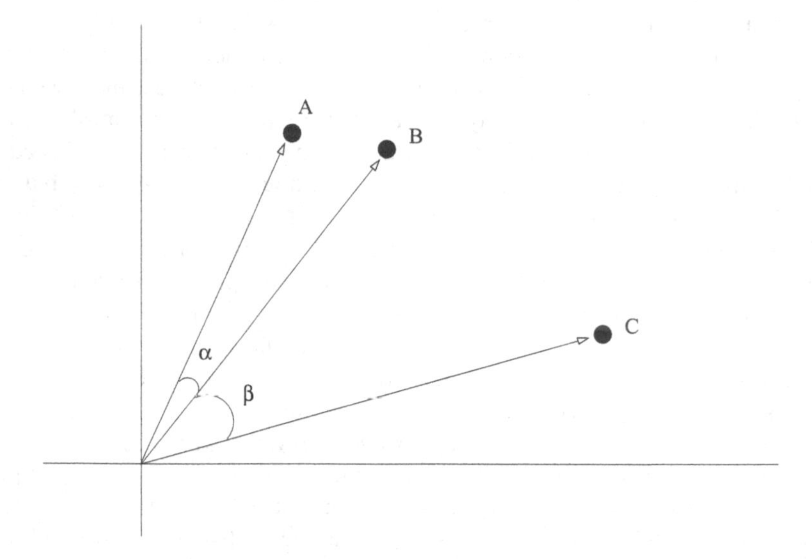

Fig. 4.2 Distance between A and B this time as cosine of angle α; distance between B and C as cosine of angle β

angle is 180° (so the cosine is -1). The cosine is calculated as

$$\frac{\Sigma_i^n (x.A_i y.A_i)}{\sqrt{\Sigma_i^n (x.A_i)^2}\sqrt{\Sigma_i^n (y.A_i)^2}}.$$

This distance ranges from -1 (meaning exactly opposite) to 1 (meaning exactly the same); 0 means 'orthogonal' or non-correlated. It is a popular distance used in several contexts, as we will see.

To implement any distance in SQL, the general pattern is: given a dataset `Data(id,dim1,dim2,...)`,

```
SELECT D1.id, D2.id, distance(D1.*, D2.*)
FROM Data D1, Data D2;
```

where `distance(D1.*, D2.*)` is the calculation of the chosen distance function using the attributes in the schema (or whatever attributes we deem relevant). For instance, the Euclidean distance over attributes `dim1, dim2, ...` becomes

```
SELECT  sqrt(pow(D1.dim1 - D2.dim1, 2) +
             pow(D1.dim2 - D2.dim2, 2) + ...)
FROM Data D1, Data D2;
```

Observe that the FROM clause generates a Cartesian product; this allows us to compare each point to each other point. However, it also creates a problem: in a dataset of size n (that is, with n data points), the number of comparisons is n^2. This is too much for even medium-sized datasets (set $n = 100,000$). It is possible to cut comparisons in half by adding

```
WHERE D1.id <= D2.id;
```

because of symmetry of distances. However, this is just a bit of relief. In Sect. 5.3 we will see a more efficient way to compute some simple distances.

Exercise 4.9 Give an SQL query to compute the Manhattan distance over a generic `Data` table.

Exercise 4.10 Give an SQL query to compute the cosine distance over a generic `Data` table.

4.3.1.1 K-Means Clustering

Clustering algorithms use distances between data points to group them together into clusters, or set of points such that the distance between any two points in a cluster is smaller than the distance between two points in any two clusters. Several strategies can be used to accomplish this. The *k-means clustering algorithm* uses the following approach:

1. Fix the number of clusters, k.
2. Pick k random data points and create the clusters by putting each point alone in a cluster. Set the *mean* (also called the *centroid*) of each cluster to be this (unique) point.
3. For each data point p, calculate the distance between p and each one of the cluster centroids; assign p to the cluster with the shortest distance.
4. After this is done, on each cluster recompute the *mean* or *centroid*.
5. Repeat the assignment of data points p to clusters by again computing the distance of p to each cluster, using the new centroid.
6. Repeat the last two steps until the clusters do not change, or until each cluster is *cohesive* enough, or for a fixed number of iterations.

The *cohesion* of a cluster can be measured in a number of ways; typical ones include taking the average of all distances between pairs of points in the cluster—sometimes, the maximum or the minimum is used instead of the average.

Clearly, this is an *iterative* algorithm, and the only way to implement it in SQL is with recursion. Recursion in SQL was explained in Sect. 4.6; here we explain the elements needed to write the final query. The following tables are needed, besides the table `Data(point-id,...)` storing the data:

- A table to store the means/centroids for each one of the clusters. We will also add a number to point out at which iteration the means were computed, since they will change over time. Thus, we have table `centroids(iteration, cluster-id, mean)`, which will be initialized with tuples `(1, 1, point)`, `(1, 2, point2), ...(1, k, pointk)`, where `point1, ..., pointk` are k randomly chosen data points.
- A table to keep track of the clustering, that is, the assignment of a cluster to each point. As before, we will use an iteration number to control the updates. Table `Clustering(iteration, point-id, cluster-id)` is sufficient. At initialization, all points in the dataset are included with an iteration number of 0 and a cluster id value of null.
- As an auxiliary table, we need to compute the distance between each point and each cluster, in order to establish to which cluster a point will go. A table `Distances(iteration, point-id, cluster-id, distance)` is not strictly necessary, but it will simplify the computation. This table can be started empty.

At each step, we need to compute a new assignment for each data point using the current mean (the one computed at the latest iteration) by updating the distances and updating the cluster centroids in turn. Thus, we have

```
INSERT INTO Distances
SELECT C.iter+1, D.point-id, C.cluster-id,
       d(D.attributes, C.mean)
FROM Data D, Centroids C
WHERE iter = (SELECT max(iter) FROM Centroids);
```

where d(...) is whatever distance function we have decided to use, and we restrict ourselves the most recent means by selecting the latest (highest) iteration—note also that we increase this latest iteration by one when inserting into the table, to signify that we are currently executing another iteration of the algorithm. After this is done, the new assignment of points to clusters can be done:

```
INSERT INTO Clustering
SELECT C.iter+1, D.point-id, D.cluster-id
FROM Distances D, Clustering C
WHERE  D.point-id = C.point-id
 and distance = (SELECT min(distance)
                 FROM Distances D2
                 WHERE D2.point-id = D.point-id)
 and iter = (SELECT max(iter) FROM C);
```

where we assign each point to the closest (minimum distance to mean) cluster. Finally, the centroids can then be recomputed:

```
INSERT INTO TABLE Centroids
 SELECT iter+1, cluster-id, value
 FROM (SELECT cluster-id, avg(dist) as value
       FROM Distances D, Clustering C
       WHERE D.cluster-id = C.cluster-id
       GROUP BY cluster-id) as TEMP,
       Centroids as C
 WHERE C.cluster-id = TEMP.cluster-id);
```

The iteration can be controlled by making sure that the iteration number does not exceed a prefixed value; this makes sure that the computation ends—something of importance in a database system!

4.3.2 The kNN Algorithm

k Nearest Neighbors is a simple and powerful algorithm, also based on using distances. Given a dataset D, a distance d on it, and a new data point p, the algorithm finds the k closest points (shortest distance) to p in D according to d. These points are the *neighbors* of p. Usually, k is set to be a small number, from 3 to 10, although this can change with datasets. Once this is done, kNN can be used for:

- Classification: if points are labeled with a class, a new point p can be assigned the class of its 'closest' (according to d) neighbor, or the most frequent class among p's k closest neighbors.
- Prediction: if an attribute y needs to be predicted from $A_1, \ldots, A_n$ on a new data point p, we can base this prediction on the y values of the k 'closest' (according to d) neighbors of p.

For classification, given dataset Data(point, class) and new point pt, we can compute pt's class as follows (using the rule of selecting the majority's class):

```
WITH kNN AS (SELECT point, class
             FROM Data
             ORDER BY distance(point, pt)
             LIMIT k)
SELECT class
FROM (SELECT class, count(*) as freq
      FROM kNN
      GROUP BY class) as classFreq
WHERE freq = (SELECT max(freq) FROM classFreq);
```

We can use weighed counts for the class or use other aggregates.

Exercise 4.11 Write the kNN algorithm using the distance of neighbor p' as the weight of p' in the computation for the majority class.

For prediction, given dataset `Data(X1,...,Xn,Y)` and new data point `(X1',...,Xn')`, we can predict the `Y` for the new data point as follows:

```
SELECT avg(Y)
FROM (SELECT Y
      FROM Data
      ORDER BY distance(X1,..,Xn,X1',...,Xn')
      LIMIT K) as KNN;
```

Again, we can weigh the average by distance or use another aggregate.

Exercise 4.12 Write the kNN algorithm for prediction using the distance of neighbor p' as the weight of p' in the computation for the majority class.

Example: Using kNN Algorithm for Prediction

Assume table `Data(x,y,...)` with both attributes numerical. Assume we want to predict the value of `y` when `x=6.5`, but there is no point in the table with that value of `x`. We use kNN with $k = 2$, and the Manhattan distance on `x`.

```
SELECT avg(y)
FROM
 (SELECT y FROM Data ORDER BY abs(6.5-x) LIMIT 2) as KNN;
```

The value of the parameter k is set per problem and it reflects a trade-off. Low ks are sensitive to outliers; larger ks are robust, but more expensive to compute. The kNN algorithm, like all algorithms based on distance, requires that dimensions be standardized so that larger numerical values do not dominate. Also, note that all attributes used to compute the distance must be numerical. Categorical variables can be handled by creating dummy variables.

4.3.3 Association Rules

Association rules are a way to look at relations between *values* of an attribute. In some datasets, we have events that involve a set of items; in databases, such sets are usually called *transactions*.[5] The idea is to determine whether there are connections among the components of transactions; in particular, to determine whether certain components appear frequently together.

Association rules first appeared in the analysis of 'market baskets,' that is, of retail transactions. In this context, an *itemset* (i.e. a set of items) refers to the products involved in a transaction, as each transaction represents a customer purchasing several products at once—here, the event is the shopping transaction, and the components are the products bought. For instance, a SALES table could have schema *(transaction-id, product, quantity, price)* to specify that in a certain transaction (identified by its id) consisted of the purchase of two (amount) loafs of French bread (product) at \$1.50 each (price); three cans of lentil soup, at \$2.50 each; and one jug of milk for \$3.00. A transaction would be represented by listing the components (product, in this case) row by row:

Transaction-id	Product	Quantity	Price
1	"French bread"	2	1.50
1	"Lentil soup"	3	2.50
1	"Milk"	1	3

The goal is to analyze the products purchased within transactions (items in the itemset) to see if this reveals any interesting pattern. Even though association rules found their first application in market analysis, they can be used in any scenario where a 'transaction' event can be identified involving several 'items.'

An association rule of the form $X \rightarrow Y$, where X and Y are itemsets, tells us that 'transactions' that contain values X are likely to also contain values Y' (i.e. $\{bread, soup\} \rightarrow \{milk\}$). X is called the *left-hand side* of the rule and Y is called the *right-hand side*. It is difficult to search for association rules because there are many possible combinations of values in a dataset (in our example, there are many items that can be purchased from your average supermarket); enumerating all combinations is tremendously costly. However, we are only interested on certain associations. In particular, we want the association to be frequent in the dataset, that is, to have a large presence, so that the chances of it being just an accident are small. Also, we want the association to be strong so that, again, we are not seeing something due merely to chance. The *support* of $X \rightarrow Y$ (in symbols, $|X, Y|$) is defined as the set of items (rows) that contain both values X and Y (in our example, the number of transactions that contain bread, soup, and milk among the products purchased). We want to focus on itemsets with high support (if such transactions are

[5] In this subsection, we use the term 'item' to refer to product or, in general, aspect of a transaction, as this terminology is deeply entrenched (i.e. see the idea of 'itemset' later).

infrequent, any pattern we find in them may be due to noise, or not very relevant). Among these, we want those where the association is strong. The *confidence* of $X \rightarrow Y$ is $\frac{|X|}{|X,Y|}$, that is, the fraction of 'transactions' that contain X among those that contain X and Y (in our example, we count which percentage contain milk and divide this by the number of transactions that involve bread, soup, and milk. This establishes how often the association is true, as a measure of the strength of the relationship among items.

Example: Association Rules

Using the table `SALES(transaction-id, product, quantity, price)`, we compute rules involving products by counting, for each pair of products A and B, the support and confidence of rule $A \rightarrow B$. For this, we first have to produce the pairs (A,B) of products in the same transaction; we do this computation first and produce a temporary table `Pairs`.

```
CREATE TABLE Pairs AS
SELECT  transaction-id, LeftHand.product as Left,
                        RightHand.Product as Right
FROM (SELECT DISTINCT transaction-id, product FROM SALES)
     AS LeftHand,
     (SELECT DISTINCT transaction-id, product FROM SALES)
     AS RightHand,
WHERE LeftHand.transaction-id = RightHand.transaction-id and
     LeftHand.product <> RightHand.product;
```

We now can count support and confidence easily; note that this is done over the original ORDERS table, since we need to include all data, even transactions with just a single product, which did not make it into PAIRS. We then join these results with the previous one:

```
SELECT Left, Right, both / (total * 1.0) as support
                    both / (countLeft * 1.0) as confidence
FROM (SELECT Left, Right, count(*) as both
      FROM Pairs
      GROUP BY Left, Right) AS Supports,
     (SELECT product, count(*) as countLeft
      FROM ORDERS
      GROUP BY product) as LeftCounts,
     (SELECT count(DISTINCT transaction-id) as total
      FROM ORDERS) AS all
WHERE Pairs.Left = Supports.Left and
      Pairs.Right = Supports.Right and
      Pairs.Left = countLeft.product
GROUP BY Pairs.Left, Pairs.Right;
```

Note that the support is expressed as a percentage (i.e. how many, out of all transactions, contain both items) since this is usually much more informative than the raw number. Note also that this will list all possible pairs, which can be very numerous. It is traditional to demand a minimal support (and sometimes, a minimal confidence also). This threshold can be added easily to the query by adding a condition with a HAVING clause:

```
HAVING support > .1;
```

and likewise for a threshold on confidence.

Note that a rule $X \rightarrow Y$ and a rule $Y \rightarrow X$ are different: clearly, they both have the same support, but they may have very different confidences. When it is not known which way an association works, calculating both possibilities and picking out the one with the higher confidence is a good idea.

This still leaves open the fact that X and Y are sets of values: we could have rules like $\{beer, bread\} \rightarrow \{diapers, milk, eggs\}$. However, there are too many sets of values for us to calculate confidence and support for each pair of sets. Each element of the set involves a self-join of SALES with itself, so a rule like $\{beer, bread\} \rightarrow \{diapers, milk, eggs\}$ would require us to join SALES with itself 4 times to have lists of 5 elements, which then we need to break into two sides (left and right), which in itself can be done in several different ways (from only one element on the left and 4 on the side to the other way around). The challenge then is to extend the approach to more than one value per side.

A nice property of support is that a set X can only have high support if each $Z \subset X$ also has high support. In our example, for instance, the set $\{beer, bread\}$ can only have high support if both $\{beer\}$ and $\{bread\}$ have high support—since the support of $\{beer, bread\}$ is at most the smallest of the supports of $\{beer\}$ and $\{bread\}$. This is the idea behind the *a priori* algorithm: first, check the support of individual values and discard those below an appropriate threshold. Next, build sets of pairs of attributes using only those that survived the filter, compute the support for these pairs of attributes, and again filter those with low support (since even if each value individually has high support, the pair may not, as they may appear in very few common transactions). The process then continues: merge pairs of attributes with high support to create a three-element set and check the support of the result, discarding those results below the threshold. At each step, sets with low support can be thrown away since they cannot 'grow' into high support sets. For instance, the support of $\{diapers\}$ is greater or equal to the support of $\{diapers, milk\}$, which is greater than or equal to the support of $\{diapers, milk, egg\}$. Support goes down as the set grows, so we eventually run out of sets to consider; depending on the dataset, this may happen relatively early.

This suggests an improvement to the strategy above: start by computing the support of individual items and then compute support and confidence only for rules involving individual items with support above a threshold. In large databases, this can make quite a bit of difference.

Example: Efficient Association Rules

In our previous example, we computed all possible pairs of items in table Pairs. Instead, we can start with

```
CREATE TABLE Candidates AS
SELECT product
FROM SALES
GROUP BY product
HAVING count(*) > threshold;
```

and run query Pairs over table Candidates joined with Sales (we need to make sure that pairs of products are in the same transaction, using `transaction-id`).

```
CREATE TABLE Pairs AS
SELECT  transaction-id, LeftHand.product as Left,
                        RightHand.Product as Right
FROM (SELECT product as Left FROM Candidates) as C1,
     (SELECT product as Right FROM Candidates) as C2,
     (SELECT DISTINCT transaction-id, product FROM SALES)
      AS LeftHand,
     (SELECT DISTINCT transaction-id, product FROM SALES)
      AS RightHand,
WHERE LeftHand.transaction-id = RightHand.transaction-id and
     LeftHand.product <> RightHand.product and
     LeftHand.product = C1.product and
     RightHand.product = C2.product;
```

Even though this query is more complex than the previous one, the table Candidates may be considerably smaller than Sales (depending on the threshold and the dataset), so this result will also be smaller than before.

Also, note that the computation of support may be pushed to the table Pairs, and only those pairs of attributes with support above the threshold are then left to compute confidence.

Exercise 4.13 Rewrite the SQL for the creation of table `Pairs` keeping only pairs of attributes with support greater than 0.2.

Using this idea, we can compute larger sets of items by using only those in Candidates and joining such items with itemsets that are themselves large. For instance, in the example above, we can join Candidates and Pairs to generate 3-element sets. This is a simplification of *A priori* but is still more efficient than an exhaustive consideration of all items.

Example: Complex Association Rules

As stated, we can generate triplets of items using the Candidates and Pairs from earlier:

```
CREATE TABLE Triplets AS
SELECT S1.transaction-id, product1, product2, product3
FROM (SELECT product as product1 FROM Candidate)
     AS C,
     (SELECT Left as product2, Right as product3 FROM Pairs)
     AS P,
     (SELECT DISTINCT transaction-id, product FROM SALES)
     AS S1,
     (SELECT DISTINCT transaction-id, product FROM SALES)
     AS S2,
     (SELECT DISTINCT transaction-id, product FROM SALES)
     AS S3,
WHERE S1.transaction-id = S2.transaction-id and
      S2.transaction-id = S3.transaction-id and
      S1.product = product1 and
      S2.product = product2 and
      S3.product = product3;
```

As stated above, this table can be further trimmed by checking for appropriate support, i.e. counting the number of transactions with all 3 products.

Exercise 4.14 Rewrite the SQL to create table Triplets by trimming all triplets with a support lower than 0.1.

But we still face two problems: first, even if (A,B) and (C) have good support, this does not guarantee that (A,B,C) has good support, so we need to check this. Second, even if (A,B,C) has good support, there are six rules that can be generated from this set: $A \rightarrow B, C$; $B \rightarrow A, C$; $C \rightarrow A, B$; $A, B \rightarrow C$; $A, C \rightarrow B$; $B, C \rightarrow A$. We can, of course, generate all of them and check their confidence (they all share the same support, which we just checked). However, going beyond 3 attributes quickly becomes a combinatorial nightmare.

Exercise 4.15 Write SQL queries using table Triplets to compute, for a triple (A,B,C), confidence for rules $A \rightarrow B, C$; $B \rightarrow A, C$; and $C \rightarrow A, B$, and keep the rule with the highest confidence.

Even using the *A priori* algorithm in its full generality has a very high cost, as it generates a large number of sets that are later disregarded. Therefore, it is customary to investigate rules involving only small sets (2 or 3 items on each set, left and right). The good news is that, for the reasons just seen, very large sets are highly unlikely to have large supports.

4.4 Dealing with JSON/XML

As we saw in Sect. 2.3.1, there are two ways of dealing with semistructured data in databases. The first one was flattening, that is, transforming the hierarchical structure into a 'flat' one that leaves the data in a table. After flattening, all the algorithms that we have seen so far apply to this data too. The second method is to create a table with a column of type XML or JSON and store the data there in one of these formats. Unfortunately, there are very few algorithms that deal with data in this format. Hence, the typical approach is to flatten XML or JSON data into a tabular format and analyze the resulting table.

Most systems have functions that allow them to extract data from an XML or JSON column and represent it as a (plain) table; unfortunately, the names and types of such functions can vary considerably from system to system, as not all of them follow the SQL standard faithfully. The basic ideas behind these (and other) functions are always the same:

- If a value a associated with several values $b_1, \ldots, b_n$ (as a sub-element, in XML; as an array, in JSON), this is flattened into tuples $(a, b_1), \ldots, (a, b_n)$. If several levels exist, the process is repeated for each level.
- If we want to retrieve part of a complex element or object, a *path* is indicated to locate the part of interest. A path represents the location of a part by giving directions to 'navigate' the complex element or object from its root.[6] In XML notation, paths are indicated with forward slashes (as in "element/subelement/...") and optionally conditions or functions in square brackets (for instance, the value of an XML attribute is accessed using '[@attribute_name]').[7] In JSON, a path is built by using the dot notation ('.') to denote attributes inside an object and the square brackets ('[]') to denote an element inside an array.
- The schema of the target table (i.e. the target to be created) is sometimes given implicitly to the function used (some functions take parameters that indicate the names and type of attributes to be created), and sometimes given explicitly (in a separate clause). To accommodate the irregular nature of XML and JSON data, attributes not mentioned (explicitly or implicitly) but present in the data are ignored; and attributed mentioned (explicitly or implicitly) but not present in the data are given a default value (by the user or by the system).

In the following examples, we illustrate these ideas by showing some of the basic functionality in Postgres and MySQL.

[6] Recall that all hierarchical data can be seen as a 'tree,' as described in Sect. 1.2; a path means simply a description of the 'route' from the root to the given element—which in a tree is always unique.

[7] There is a whole language, XPath, devoted to denoting paths in XML, as they can become quite complex expressions. We do not discuss it here.

In Postgres, the main function for flattening XML data is called *xmltable*. Its (simplified) format is

```
xmltable(
   row_expression PASSING document_expression
       COLUMNS name { type [PATH column_expression]
                           [DEFAULT default_expression]
                           [NOT NULL | NULL]
                           | FOR ORDINALITY })
```

The xmltable function produces a table based on arguments:

- `document_expression`, which provides the XML document to operate on. The argument must be a well-formed XML document; fragments/forests are not accepted.
- `row_expression`, which is an XPath expression that is evaluated against the supplied XML document to obtain an ordered sequence of XML nodes. This sequence is what `xmltable` transforms into output rows.
- An optional set of column definitions, specifying the schema of the output table (if the COLUMNS clause is omitted, the rows in the result set contain a single column of type xml containing the data matched by `row_expression`). If COLUMNS is specified, each entry gives a single column name and type (other clauses are optional). A column marked FOR ORDINALITY will be populated with row numbers matching the order in which the output rows appeared in the original input XML document. At most one column may be marked FOR ORDINALITY. The `column_expression` for a column is an XPath expression that is evaluated for each row, relative to the result of the `row_expression`, to find the value of the column.

Note that in XML not all elements may have all attributes; this will result in a table with nulls unless a DEFAULT value is specified.

Example: XML Flattening in Postgres

Suppose we have the following data in XML:[8]

```
CREATE TABLE xmldata AS SELECT
xml $$
<ROWS>
  <ROW id="1">
    <COUNTRY_ID>AU</COUNTRY_ID>
    <COUNTRY_NAME>Australia</COUNTRY_NAME>
  </ROW>
  <ROW id="5">
    <COUNTRY_ID>JP</COUNTRY_ID>
    <COUNTRY_NAME>Japan</COUNTRY_NAME>
    <PREMIER_NAME>Shinzo Abe</PREMIER_NAME>
```

[8]This example is taken from the Postgres documentation.

```
    <SIZE unit="sq_mi">145935</SIZE>
  </ROW>
  <ROW id="6">
    <COUNTRY_ID>SG</COUNTRY_ID>
    <COUNTRY_NAME>Singapore</COUNTRY_NAME>
    <SIZE unit="sq_km">697</SIZE>
  </ROW>
</ROWS>
$$ AS data;
```

Note that the above creates a table called `xmldata` with a single attribute called `data`, of type XML, on it.

```
SELECT xmltable.*
FROM xmldata,
     XMLTABLE('//ROWS/ROW' PASSING data
              COLUMNS id int PATH '@id',
                      order FOR ORDINALITY,
                      "COUNTRY_NAME" text,
                      ct_id text PATH 'COUNTRY_ID',
                      szsqkm float
                           PATH 'SIZE[@unit = "sq_km"]',
                      size_other text
            PATH 'concat(SIZE[@unit = "sq_km"], " ",
                         SIZE[@unit != "sq_km"]/@unit),
                    premier_name text PATH 'PREMIER_NAME'
                           DEFAULT 'not specified');
```

The above query transforms the data in `xmldata` into the following table:

```
id|order|COUNTRY_NAME|ct_id|szsqkm|size_other |premier_name
--+-----+------------+----------+----------+----------+------
1 |    1 | Australia  | AU  |      |             | not specified
5 |    2 | Japan      | JP  |      |145935 sq_mi| Shinzo Abe
6 |    3 | Singapore  | SG  |  697 |             | not specified
```

Note how an attribute in the XML data (`SIZE`) has been split into two attributes in the table, depending on the value of XML attribute `unit`. Note also that there are missing values that are not explicitly marked (in attributes `size_sq_km` and `size_other`) and missing values explicitly marked (in attribute `premier_name`).

Exercise 4.16 The above practice of dealing with missing data in several ways is unwise. Modify the query above so that all absent data is marked by the string 'NA.'

As for JSON data, Postgres provides a set of functions that can be combined to flatten a JSON collection into a table:

- `JSON_each(JSON)` expands the outermost JSON object into a set of key/value pairs; each pair becomes a row in the resulting table.

```
SELECT *
FROM JSON_each('{"a":"foo", "b":"bar"}')

 key | value
-----+-------
 a   | "foo"
 b   | "bar"
```

- To find values inside a complex object, `JSON_extract_path(from_JSON JSON, VARIADIC path_elems text[])` returns the JSON value found following the argument `path_elems`:

```
JSON_extract_path('{"f2":{"f3":1},
                    "f4":{"f5":99,"f6":"foo"}}','f4')

{"f5":99,"f6":"foo"}
```

  Here ‘f4’ is the path.
- `JSON_populate_recordset(base anyelement, from_JSON JSON)` expands the outermost array of objects in the second argument to a set of rows whose columns match the record type defined by the first argument.

```
SELECT *
FROM JSON_populate_recordset(null::myrowtype,
                         '[{"a":1,"b":2},{"a":3,"b":4}]')

 a | b
---+---
 1 | 2
 3 | 4
```

- `JSON_to_recordset(JSON)` builds a table (set of records) from a JSON array of objects. The schema of the table (i.e. the structure of the record) is defined with AS clause.

```
SELECT *
FROM JSON_to_recordset('[{"a":1,"b":"foo"},
                        {"a":"2","c":"bar"}]')
  as x(a int, b text);

 a |  b
---+-----
 1 | foo
 2 |
```

  Note that elements not mentioned in the AS clause are ignored; elements mentioned but not present in the JSON data have an empty string to denote the missing value.

In MySQL, the function to deal with XML is called `ExtractValue`, and it takes a JSON object and a path into the object, it extracts the value of the path in the object. To break down a JSON object into parts, we call `ExtractValue` repeatedly with the same object and the paths leading to the different parts.

```
SELECT ExtractValue('<a>ccc<b>ddd</b></a>', '/a') AS val1,
       ExtractValue('<a>ccc<b>ddd</b></a>', '/a/b') AS val2,
       ExtractValue('<a>ccc<b>ddd</b></a>', '//b') AS val3,
       ExtractValue('<a>ccc<b>ddd</b></a>', '/b') AS val4,
       ExtractValue('<a>ccc<b>ddd</b><b>eee</b></a>', '//b')
         AS val5;

+------+------+------+------+---------+
| val1 | val2 | val3 | val4 | val5    |
+------+------+------+------+---------+
| ccc  | ddd  | ddd  |      | ddd eee |
+------+------+------+------+---------+
```

As it can be seen in the previous examples, the process is laborious due to the need to specify parts of the XML/JSON objects by giving paths into the parts of the object that are of interest. Unfortunately, there is no work-around this—but since most Data Mining and Machine Learning tools will not work directly with XML or JSON data, one must become familiar with these functions in order to transform the data as appropriate.

4.5 Text Analysis

The last type of data is unstructured, or text, data. There are, roughly speaking, three levels of text analysis:

- *Information Retrieval (IR)*: IR sees documents as *bags of words*: the semantics of a document are characterized by the words it contains. IR systems support *keyword search*, the retrieval of some documents in a collection by using a list of keywords. Documents containing those keywords are retrieved. The idea is that the user will pick keywords that documents of interest are likely to use. The retrieved documents are *ranked* to signify how relevant the documents are for the given keywords. This is the technique behind web search, with each web page is seen as a document. Most relational databases nowadays support keyword search, and this is the focus of this section. Even though this is the simplest form of text analysis, other tasks like *sentiment analysis* can be performed using IR techniques.
- *Information Extraction (IE)*: IE tries to extract snippets of information from text; such snippets are usually described by rows in certain tables. The schema of the tables depends on what information can be obtained from the text. For instance, the sentence "Paris is the largest city in France, and also its capital" can result in an entry *(Paris, France)* in a table with schema *(capital, country)*. Note that not all information presented in the sentence is extracted. However, what is extracted is now in a database-like form and can be analyzed with SQL queries. The extraction relies on a number of techniques, from simple pattern matching to neural networks. IE is usually not implemented in databases due to its complexity.

- *Natural Language Processing (NLP)*: NLP analyzes both the syntax and semantics of each sentence. A *parser* breaks a sentence into its components, and a semantic analyzer then uses this result to extract all the information from the sentence. For instance, the sentence of our previous example would be parsed as a complex sentence that can be seen as the conjunction of two simple sentences: "Paris is the largest city in France," and "Paris is the capital of France." Within each sentence, "Paris" is determined to be the subject, "is" the main verb, and what follows it is a direct complement that can in turn be broken into parts. NLP goes 'deeper' into analysis than IE: besides getting more information, the information retrieved is usually represented in a richer format (typically, the example would be expressed with a logic formula stating that for all cities x of France, if y is the population of x, then the population of Paris is greater than or equal to y). NLP analysis can get very complex and is usually carried out using deep learning techniques. As a result, NLP is usually also not implemented in databases.

In summary, IE and NLP are complex, specialized fields that rely heavily on machine learning; IR, while simpler, provides some basic tools that can be profitably used for analysis. We explain here how IR can be carried out in a database.

IR sees documents as *bags of words*; the semantics of a document can be characterized by its *content words*. *Non-content words* or *stopwords* (also called *function words* [1, 3, 12]) are words that carry no informational content (usually, articles and prepositions: *a, the, of*). They are present in almost any document, so they have no discriminatory value.[9] If we eliminate them, we are left with the content words. Note that no syntax or semantics is used; only word appearance is important. Not even order is used: `the cat is on the mat` and `the mat is on the cat` are exactly the same in IR (except for indices that keep word offsets, see below). Beyond the order of words in a sentence, we also throw away relationships among sentences, and any logical structure in the document (to build arguments, etc.).

Documents are *tokenized*, i.e. divided into discrete units or tokens. This is accomplished through a series of steps.

- *Determining canonical words*: words that are strongly related may be converted to a common term. For instance, verb forms (past, present, tense) may all be converted to a root. A particular example of this is *stemming*, getting rid of word inflections (prefixes, suffixes) to link several words to a common root (for instance, transforming *running* to *run*). Common stemming methods are based on morphological knowledge (and hence are language dependent).[10] Note that stemming introduces some risk: Porter's algorithm, one of the best known stemmers for English, stems *university* and *universal* to *univers*. Some stemmers also transform the case of letters (all to lowercase or uppercase).

[9] Most systems provide a list of stopwords, also sometimes called a *negative dictionary*.

[10] Obviously this only applies to languages where words can be declined.

- Another improvement is to accept *phrases*. Technically, phrases are n-grams (i.e. n terms adjacent in the text) that together have a meaning different from the separate terms, e.g. *operating system*. In systems that accept them, phrases are terms of their own right. However, to discover phrases may be complicated. There are basically two approaches, syntactical and statistical. In the syntactic approach, allowed combinations are listed by syntactic categories, i.e. *noun* + *noun*. However, this approach is weak, as most rules allow non-phrases. The statistical approach consists of looking at the number of occurrences of n terms $t_1, t_2, \ldots, t_n$ together and determining if this number is higher than it could be expected if the terms were independent (i.e. higher than the product of their individual occurrences). Recognizing phrases in general is a complex problem; in a language like English, phrases can become quite large and complex (*simulated back-propagation neural network*, *apples and oranges*).
- *Approximate string matching* is required in order to deal with typos, misspellings, etc., which can be quite frequent in some environments (e.g. the web) due to the lack of editorial control. There are two ways to attack the problem: one is by using methods like the ones shown in Sect. 3.3.1.2. The other approach is to break down each word into n-grams, or sequences of n characters, and compare the overlap of sequences between two given words. The parameter n depends on the language characteristics, like typical syllable size. Because approximate term matching may be expensive, it is typically not used by default in any IR approach.
- Another technique is to have a *thesaurus* or a similar resource to catch *synonyms* and choose a term to stand for all synonyms. This reduces the number of terms to deal with and allows the user to denote the same concept with different terms. However, thesaurus are usually built by hand and therefore their quality and coverage may vary substantially.

To explain how databases support IR, we introduce some terminology. Let D be a collection of m documents, that is, $|D| = m$ (D is sometimes called a *corpus*). Let T be the collection of all terms in D; for any $t \in T$, we denote by D_t the set of documents where t appears. For term t and document d, we denote with $tf(t, d)$ the occurrence frequency of t in d. The intuition behind this number is that if the term occurs frequently in a document, then it is likely to be very significant (and vice versa: if a term is only mentioned once or twice, it may be a mention in passing, meaning that the term does not represent the contents of the document). Note that since tf is an absolute value (not normalized) we may need to normalize it: as is, it tends to favor larger documents over short ones. tf can be normalized by the sum of term counts:

$$tf(t, d) = \frac{tf(t, d)}{\sum_{d' \in D} tf(t, d')}$$

or by the largest sum:

$$tf(t,d) = \frac{tf(t,d)}{max_{d' \in D} tf(t,d')}.$$

The *inverse document frequency (idf)* of t is computed as $idf(t, D) = \log \frac{|D|}{|D_t|} = \log \frac{m}{n}$ (note: when $n = m$, idf is 0; when $n = 1$, idf will be as large as possible). The idf tries to account for the fact that terms occurring in many documents are not good discriminators. The number $|D|$ acts as a normalization factor; we could also use $max_{t' \in T} D_{t'}$. A very commonly used normalization factor is $|D| - |D_t|$, i.e. the number of documents *not* containing the term.

Note that this is a property of the corpus a whole, not just of a document!

The *term frequency (tf)* of a term is the number of documents where the term occurs, i.e. $|D_t|$. In general, terms with high document frequencies are poor at discriminating among documents, since their appearance may not be significant.[11] However, terms that appear very rarely are also of limited help, as they do not tell us much about the corpus as a whole. It has been found that the best terms for searching are those that have medium document frequencies (not too high, not too low) (also, among terms occurring on the same number of documents, those with a higher variance are better).

A *tf-idf weight* is a weight assigned to a term in a document, obtained by combining the tf and the idf. Simple multiplication can be used, especially with the normalized tf:

$$TFIDF(t,d) = TF(t,d) \times idf(t,D).$$

To compute this weight, most systems create an *inverted (full text) index*. This is a list of all words in D and, for each word, a list of the documents where they appear. An inverted index could be represented by a table with schema `WORDS(term, docid)`. With a table like this, tf and idf can be computed in SQL:

```
SELECT term, docid, count(*)/len as tf
FROM WORDS,
     (SELECT  docid, count(*) as len
      FROM WORDS
      GROUP BY docid) as Temp
WHERE WORDS.docid = Temp.docid
GROUP BY term, docid;


SELECT term, total/count(distinct docid) as idf
FROM WORDS,
     (SELECT count(distinct docid) as total FROM WORDS)
      as Temp
GROUP BY term;
```

[11] Recall that ubiquitous presence was the reason to get rid of stopwords, but stopwords have high frequency too.

Most systems compute tf and idf from the index and use it for ranking results when keyword search is used, as explained next.

Keyword search consists of searching, among a corpus of documents, for the ones relevant for a certain goal; the search is based on a list of words, called *keywords*, that we expect to appear in any such document. As stated, most database systems support keyword search. If a table is created with at least one attribute of type `Text`, this attribute is considered to contain a corpus, with each row's value for the attribute being a document. It is possible to create an inverted index in such an attribute and then carry out keyword search on it. We now describe how keyword search is supported in MySQL and Postgres.

MySQL comes with its own list of stopwords, but it can be overwritten by a user, as follows: when using the `InnoDB` engine,

```
CREATE TABLE my_stopwords(value VARCHAR(30));
Query OK, 0 rows affected (0.01 sec)

INSERT INTO my_stopwords(value) VALUES ('or');
Query OK, 1 row affected (0.00 sec)

SET GLOBAL innodb_ft_server_stopword_table = 'my_stopwords';
Query OK, 0 rows affected (0.00 sec)
```

With the `MyISAM` engine, one creates a stopword file and then calls MySQL with option:

`-ft-stopword-file=file_name`

in the command line.

Keyword search in MySQL is based on the predicate

`MATCH(columns) AGAINST string`

`MATCH()` takes a comma-separated list that names the columns to be searched; `AGAINST` takes a string to search for, and an optional modifier to indicate type of search. For each row in the table, `MATCH()` returns a relevance score (based on $tf - idf$). There are three types of searches:

- Natural language search: searches for the string as is. A phrase that is enclosed within double quote (") characters matches only rows that contain the phrase literally, as it was typed. Without quotes, the system searches for the words in no particular order.
- Query Expansion search: after natural language search, words from the most relevant documents are added to the query, and the search is repeated with these additional words.
- Boolean search: the string is interpreted as a pattern, and the system searches for matches. The types of patterns allowed are described below.

A quick example shows how this is done:

```
CREATE TABLE articles (
   id INT UNSIGNED AUTO_INCREMENT NOT NULL PRIMARY KEY,
   title VARCHAR(200),
   body TEXT,
```

```
    FULLTEXT (title,body)
  ) ENGINE=InnoDB;

SELECT *
FROM articles
WHERE MATCH (title,body)
      AGAINST ('database' IN NATURAL LANGUAGE MODE);
```

MATCH() can be used in the SELECT clause and in the WHERE clause. When MATCH() is used in a WHERE clause, rows are returned automatically sorted with the highest relevance first. When used in SELECT, the score assigned to each row is retrieved, but the returned rows are not ordered. For instance, the following returns the rows scored but in no particular order:

```
SELECT id, MATCH (title,body)
FROM articles
AGAINST ('Tutorial' IN NATURAL LANGUAGE MODE) AS score
```

To get the results ordered, we can repeat the MATCH() predicate in the WHERE clause:

```
SELECT id, body, MATCH (title,body) AGAINST
    ('Security implications of running MySQL as root'
    IN NATURAL LANGUAGE MODE) AS score
FROM articles
WHERE MATCH (title,body) AGAINST
    ('Security implications of running MySQL as root'
    IN NATURAL LANGUAGE MODE);
```

We can also sort results by relevance by using an ORDER BY clause:

```
SELECT id, title, body, MATCH (title,body)
      AGAINST ('database' IN BOOLEAN MODE) AS score
FROM articles
ORDER BY score DESC;
```

For Boolean search, several operators are supported:

- + acts like AND: all words must be present. If nothing is used, OR (some words must be present) is the default.
- – acts like NOT: word must be absent.
- @distance requires that word appear within a certain distance of each other (usually a 'distance' of *k* here means '*k* words apart').
- " (quotes): literal phrase.

The following table gives examples of how Boolean search can be used[12]

'apple banana'	Find rows that contain at least one of the two words
'+apple +juice'	Find rows that contain both words
'+apple macintosh'	Find rows that contain the word "apple" but rank rows higher if they also contain "macintosh"
'+apple −macintosh'	Find rows that contain the word "apple" but not "macintosh"
'+apple macintosh'	Find rows that contain the word "apple," but if the row also contains the word "macintosh," rate it lower than if row does not
'+apple +(>turnover <strudel)'	Find rows that contain the words "apple" and "turnover," or "apple" and "strudel" (in any order), but rank "apple turnover" higher than "apple strudel."
'apple*'	Find rows that contain words such as "apple," "apples," "applesauce," or "applet."
'"some words"'	Find rows that contain the exact phrase "some words."

Postgres also has an inverted ('full text') index for documents, used to support keyword search. Like MySQL, Postgres uses a dictionary of stopwords. The text search operator in Postgres is represented by the '@@' symbol. It operates of what Postgres calls a 'tsvector' (a representation of the document, with all words normalized) and a 'tsquery' (a list of keywords representing the search criteria, also normalized). There are functions `to_tsquery`, `plainto_tsquery` and `phraseto_tsquery` that are helpful in converting user-written text into a proper tsquery (there is also `to_tsvector` for tsvectors). The tsquery may combine multiple terms using AND ('&'), OR ('|'), NOT ('!'), and FOLLOWED BY ('$<->$') operators. The AND/OR/NOT operators are interpreted differently when they are used within the arguments of the FOLLOWED BY, since within FOLLOWED BY the exact position of the match is significant. Let a, b, c be keywords:

- The tsquery '$!a$' matches only documents that do not contain a anywhere, but '$!a <-> b$' is interpreted as "no a immediately after a b (but okay somewhere else in the document)";
- The tsquery '$a\&b$' normally requires that a and b both appear somewhere in the document, but '$(a\&b) <-> c$' requires a and b to appear immediately before a c.

Example: Keyword Search in Postgres

The following are examples of keyword searches in Postgres; they all return True except the second one, which is False:

```
SELECT 'a fat cat sat on a mat and ate a fat rat'::tsvector @@
       'cat & rat'::tsquery;
```

[12]From MYSQL's documentation.

```
SELECT 'fat & cow'::tsquery @@
       'a fat cat sat on a mat and ate a fat rat'::tsvector;

SELECT to_tsvector('fat cats ate fat rats') @@
                   to_tsquery('fat & rat');

SELECT to_tsvector('fatal error') @@
                  to_tsquery('fatal <-> error');
```

Another tool of text analysis is to generate (and often, count) the number of *n-grams* in a document. An n-gram is simply a sequence of n words; the most common case is $n = 2$ (called *bigrams*) and $n = 3$ (called *trigrams*). For instance, in the sentence "Mary had a little lamb," the bigrams are "Mary had," "had a," "a little," "little lamb."

To generate bigrams, we must break down a text into a sequence of words, with each word's position in the sequence explicitly marked. That is, we want to go from "Mary had a little lamb" to a set of pairs ("Mary," 1), ("had," 2), ("a," 3), ("little," 4), ("lamb," 5). In some systems, there are functions that do this for us; when such functions are not present, the process may be quite elaborate. Here we illustrate how this can be achieved in Postgres.

Assume table `user_comments(id int, comments text)` like this:

```
 comment_id |      comments
------------+-------------------------------
          1 | "i dont think this sam i am"
          2 | "mary had a little lamb"
(2 rows)
```

The breakdown process works in 3 steps. First, we make the comments into arrays of words:

```
CREATE TABLE word_list as (
SELECT id as comment_id,
        string_to_array(
          regexp_replace(
            lower(comment),
            E'[^a-z0-9_]+', ' ', 'g'),
        ' ') as word_array
FROM user_comments);
```

First we use regexp_replace to clean up the text, converting all the characters we do not care about to spaces. The 'g' at the end tells Postgres to replace all the matches, not just the first. Then we use string_to_array with a space (' ') as its split parameter to convert the cleaned comments into arrays. At the same time we will select the id of the original comment as that will be helpful later. This creates the following result:

```
 comment_id |          word_array
------------+-------------------------------
          1 | {i,dont,think,this,sam,i,am}
          2 | {mary,had,a,little,lamb}
(2 rows)
```

Second, we break down the arrays into rows and keep the order:

```
CREATE TABLE word_indexes as (
SELECT comment_id, word_array,
       generate_subscripts(word_array, 1) as word_id
FROM word_list);
```

This uses the function `generate_subscript`, which generates a sequence 1, 2, ..., m, where m is the size of an array passed as first argument. This generates the table

```
 comment_id |          word_array          | word_id
------------+------------------------------+---------
          1 | {i,dont,think,this,sam,i,am} |       1
          1 | {i,dont,think,this,sam,i,am} |       2
          1 | {i,dont,think,this,sam,i,am} |       3
          1 | {i,dont,think,this,sam,i,am} |       4
          1 | {i,dont,think,this,sam,i,am} |       5
          1 | {i,dont,think,this,sam,i,am} |       6
          1 | {i,dont,think,this,sam,i,am} |       7
          2 | {mary,had,a,little,lamb}     |       1
          2 | {mary,had,a,little,lamb}     |       2
          2 | {mary,had,a,little,lamb}     |       3
          2 | {mary,had,a,little,lamb}     |       4
          2 | {mary,had,a,little,lamb}     |       5
```

Third, we use the array index to get individual words out, together with their position:

```
CREATE TABLE numbered_words AS
SELECT comment_id, word_array[word_id] word, word_id as pos
FROM word_indexes);
```

which yields the table

```
comment_id |  word  | pos
-----------+--------+-----
         1 | i      |   1
         1 | dont   |   2
         1 | think  |   3
         1 | this   |   4
         1 | sam    |   5
         1 | i      |   6
         1 | am     |   7
         2 | mary   |   1
         2 | had    |   2
         2 | a      |   3
         2 | little |   4
         2 | lamb   |   5
```

Now we can make bigrams by self-join this table with itself:

```
SELECT  nw1.word,  nw2.word
FROM numbered_words nw1
     join numbered_words nw2 on
        nw1.word_id = nw2.word_id - 1
        and nw1.comment_id = nw2.comment_id;
```

From here we can do further analysis; for instance, we can get the bigram frequencies:

```
SELECT nw1.word, nw2.word, count(*)
FROM numbered_words nw1
     join numbered_words nw2 on
       nw1.word_id = nw2.word_id - 1
       and nw1.comment_id = nw2.comment_id
GROUP BY nw1.word, nw2.word
ORDER BY count(*) desc;
```

Clearly, trigrams (and n-grams) in general can also be obtained using 2 (or $n - 1$) self-joins.

Another type of analysis that has become very popular with text is *sentiment analysis* (also called `sentiment detection`) [12]. Given a text (document) T and a target t (which can be a product, or a person, or an idea), we assume that T expresses some opinions about t, either in a *positive* (favorable, supportive) or *negative* (critical) way. While sentiment analysis can be quite tricky, a rough approximation can be achieved as follows: first, we come up with a list giving certain words (mostly, adjectives) a positive or negative score (for instance, 'good' or 'great' would have a positive score, while 'terrible,' 'harmful' would have a negative score). Then we create a score for each document T by adding up the scores of words in T that are in our list. Assume, for instance, that we have a table `Sentiment(word,score)`, where `score` is a number between n and $-n$, that gives the 'sentiment value' of the word. Then we can break each document into a list of words as we saw previously when dealing with bigrams; given table `Words(docid, word)` we can estimate a score per document:

```
SELECT W.docid, sum(S.score) as sentiment
FROM Sentiment S, Words W
WHERE S.word = W.word
GROUP BY W.docid;
```

However, it should be clear that this analysis is very approximate; for instance, positive words within the scope of a negation actually represent a negative sentiment ("this product was not good at all"). There are also other subtle problems, like irony. More sophisticated NLP techniques are currently used to extract sentiment; but when dealing with large collections, the above can be a good first step to focus on a smaller set of documents.

Exercise 4.17 As an improvement over the approach proposed, create a table from `Words(docid,word,position)` where all words that are preceded by an negation ("non," "no," "isn't," "wasn't," "don't") have their weight changed from m to $-m$ (note that this will turn positive words into negative words and negative words into positive ones).

4.6 Graph Analytics: Recursive Queries

Graphs can be analyzed in many different ways, but most analyses look for connectivity (what paths exist in the graph) and patterns (is there a part of the graph that has this links?), since one of the fundamental characteristics of networks (or graphs in general) is connectivity. We might want to know how to go from A to B, or how two people are connected, and we also want to know how many "hops" separate two nodes—in networks, 'distance' usually refers to the length of the shortest path between two nodes and is also called 'degree of separation.' For instance, social networks like LinkedIn show our connections or search results sorted by degree of separation, and trip planning sites show how many flights you have to take to reach your destination, usually listing direct connections first.

We start with paths: assume the standard representation of graphs with two tables called `NODES(id, ....)` and `EDGES(source, dest, weight)` as introduced in Sect. 2.3.2. Listing the nodes directly connected to a given node i (that is, connected by a path of length 1) is very simple:

```
SELECT *
FROM nodes N JOIN edges E ON N.id = E.dest
WHERE e.source = i;
```

or, in the case of undirected edges:

```
SELECT * FROM nodes WHERE id IN (
  SELECT source FROM edges WHERE dest = i
  UNION
  SELECT dest FROM edges WHERE source = i);
```

Nodes connected by a 2-step path are also easy to get:

```
SELECT E1.source, E2.dest
FROM edges E1 JOIN edges E2 ON (E1.dest = E2.source);
```

Every step requires a join. To get all nodes connected by a 3-step path, we use

```
SELECT E1.source, E3.dest
FROM edges E1 JOIN edges E2 ON (E1.dest = E2.source)
     JOIN edges E3 ON (E2.dest = E3.source);
```

Note that we are joining the EDGES table with itself to create paths, since this is how paths are expressed in this representation (see Fig. 4.3). In general, finding a path with length n requires $n - 1$ joins.

Fig. 4.3 Paths as self-joins

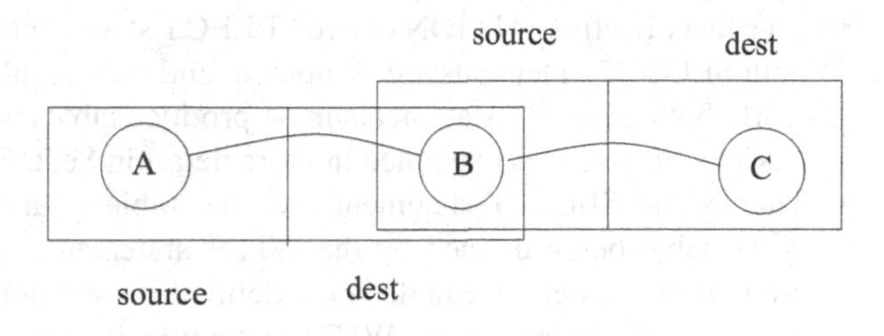

source	dest
A	B
B	C
...	...

The problem is that finding arbitrary paths requires more flexibility, since we do not know in advance how long a path may be, and therefore we cannot fix the number of steps. In order to fully analyze the graph, we need to use recursive queries.

One way to do this is to create a temporary table holding all the possible paths between two nodes. This is called the *transitive closure* of the graph and can be done in a single statement as follows:

```
WITH RECURSIVE
transitive_closure(source, dest, distance, path_string) AS
(SELECT source, dest, 1 AS distance,
        source || '.' || dest || '.' AS path_string
 FROM edges
UNION ALL
 SELECT tc.source, e.dest, tc.distance + 1,
        tc.path_string || e.dest || '.' AS path_string
FROM transitive_closure AS TC JOIN edges AS E
     ON TC.dest = E.source
WHERE TC.path_string NOT LIKE '%' || E.source || '.%')
SELECT * FROM transitive_closure
ORDER BY source, dest, distance;
```

We now describe the example in detail:

- We start with the `WITH RECURSIVE` statement. In some systems, the keyword `RECURSIVE` does not need to be used, simply `WITH` will result in a recursive query as the system detects the general pattern of such queries (explained next);

- The query itself is a UNION of two SELECT statements, which works as follows: both SELECT statements are computed, and their results put together (for this to work, both SELECT statements must produce tables with the same schema; the UNION operation is explained in more detail in Sect. 5.4).
- The second SELECT statement uses the table `transitive_closure`, which is the table being defined by the WITH statement. This is what makes this a *recursive* statement: the table being defined is used in the definition. The system computes the result of the WITH statement in stages, as follows: in the first stage, the table `transitive_closure` is created. At this point, it is an empty table (it contains no data). Hence, when the UNION statement is executed, the second query (which uses `transitive_closure` on its FROM clause, to be joined with data table `edges`) yields nothing, since the join of two tables, one of which is empty, yields an empty table. However, the first SELECT statement of the UNION, which simply uses data table `edges` can be (and is) executed. The result is that, after this first stage, the UNION picks all the results from the first SELECT statement and deposits them in table `transitive_closure`; the second SELECT statement does not contribute anything. But now the system repeats the computation: it executes the UNION statement again, but this time table `transitive_closure` has data on it (the result of the first stage). On this new computation, the first SELECT statement again grabs data from table `edges` to add to `transitive_closure`, but this is the same data that we previously added, so this is ignored. However, the second SELECT statement this time can actually be carried out and it does, taking the join of `transitive_closure` and `edges`. Whatever is produced by this second SELECT statement is now the result of the UNION and is added to `transitive_closure`. Once this is done, the system repeats the computation again. This time (all times except the very first one) the first SELECT statement in the UNION brings nothing new and so is discarded, while the second SELECT statement may (or may not) yield additional tuples. As far as the UNION adds data to `transitive_closure`, the system will keep on repeating the computation. When, at some point, the UNION yields nothing new, no data is added to `transitive_closure`, and the whole computation ends. Intuitively, the first stage adds existing edges (1-step paths) to `transitive_closure` (essentially copying `edges` in `transitive_closure`); the second stage joins `transitive_closure` with `edges` (and, since at this point `transitive_closure` is a copy of `edges`, it joins `edges` with itself); the third stage again joins `transitive_closure` with `edges`—but since now `transitive_closure` contains all 2-step paths, it produces 3-step paths. The process continues adding one more step to each path that can be extended, until we run out of paths.
- Notice that in the WHERE condition of the second SELECT there is a check that stops the recursion in the presence of loops. As we go adding more steps to `transitive_closure`, we also add a string representing the path created: the statement

```
source || '.' || dest || '.' AS path_string
```

creates an initial string with the first two nodes, separated by dots (recall that || is string concatenation), while the statement

```
tc.path_string || e.dest || '.' AS path_string
```

adds, to the existing string, a new node reached on each recursive step. The condition NOT LIKE makes sure that the node we are about to add to the path is not already in there. This is very important to avoid the system looping without end.

Once we have the transitive closure, we can find if any two arbitrary nodes are connected or not, and if so, what path(s) exist between them.

Example: Connectivity in Flights

Assume a table `Flight(src, dst, price, ...)` that lists direct flights between airport `src` and airport `dst`, together with their price and other information. Suppose a customer is interested in flying from Boston to Los Angeles. There may be direct flights or there may not, but sometimes it is actually cheaper not to fly direct, so even if direct flights exist we may want to check alternatives. We write the query

```
WITH RECURSIVE
  travel(src, dst, total_price, itinerary, num_stops) AS
(SELECT src, dst, price, src ||'-'|| dst, 0)
 FROM Flights
 WHERE src = 'BOS'
UNION ALL
 SELECT T.src, F.dst, total_price + F.price,
        T.itinerary || '-' || F.dst,
        T.num_stops + 1
 FROM travel T, Flights F
 WHERE T.dst = F.src and
       position(F.src in T.itinerary) = 0)
SELECT *
FROM travel
WHERE src = 'BOS' and dst = 'LAX';
```

Note that we only copy flights that start at Boston airport (code 'BOS') in the first stage, since it makes no sense to *start* the trip somewhere else. However, we do not stop as soon as we find the Los Angeles airport (code 'LAX'), since we may find a 3-leg flight that is cheaper than another 2-leg flight; hence we do not want to stop searching as soon as we have reached LAX in some way. However, in the end we only retrieve flights that end at 'LAX.' Note also that at each stage, the total price (which was initialized with the price of the initial leg) is increased by the price of the last leg, the itinerary (which was initialized with the source and destination of the first flight) is enlarged with the new destination, and the number of stops (which was started at 0) is increased by 1.

Exercise 4.18 Modify the query in the previous example so that it does not return flights with more than 3 legs. Apply your query to the `ny_flights` dataset to get all flights from NYC (any airport) to Los Angeles ('LAX') in 3 or fewer legs.[13]

Exercise 4.19 A more realistic example would check that the departure time of a flight is within a reasonable margin (say, 2 h) of the arrival time of the previous flight. Assume there are attributes `arrival_time` and `departure_time` in `Flights` and modify the query in the previous exercise to only add legs to the itinerary if they fulfill this condition.

As for patterns, we are usually interested in finding a subset of nodes such that they are connected in a certain way. A typical example is the search for *triangles*, sets of three nodes with each one connected to the other two. Counting triangles is a basic tool for graph analysis (used, for instance, to spot fake users in social media). The following query counts triangles in our graph:

```
SELECT e1.source, Count(*)
FROM edges E1 join edges E2 on E1.dest = E2.source
     join edges E3 on E2.dest = E3.source
     and E3.dest = E1.source and E2.source <> E3.source
GROUP BY E1.source;
```

Another common pattern is to look for nodes that are not directly connected but have many common neighbors. Suppose we have computed the transitive closure `TC(source, dest, distance)` of a social network while keeping the distances between nodes, as above. Note that we can identify all the *neighbors* of a give node—nodes that are connected directly, so they have a distance of 0 (in fact, we can determine neighborhoods of any radius [any number of steps] for any given node). We want to find pairs of nodes a, b, such that they are not neighbors of each other, but they have more than n common neighbors (note that the fact that they have common neighbors implies that a and b are at a distance of 1):

```
SELECT TC1.source, TC1.dest, count(distinct TC2.dest) as cn
FROM TC as TC1, TC as TC2, TC as TC3
WHERE TC1.distance = 1 and
      TC2.source = TC1.source and TC2.distance = 1 and
      TC3.source = TC1.dest and TC3.distance = 1 and
      TC2.dest = TC3.dest
GROUP BY TC1.source, TC1.dest
HAVING cn > n;
```

In this query, we are using renaming of the table to compare 3 copies of it: one, to make sure the nodes of interest are directly connected, and two other copies, one for the neighbors of each node, which are then compared to each other. Finally, the grouping allows us to count how many such common neighbors (`'nc'`) we have found.

---

[13]Note: this query will take some time even in a powerful PC! Make sure your database system has plenty of memory.

The idea here is first to compute the transitive closure of the graph to gather information about the connectivity in the network. We can gather distances (as above) to distinguish nodes that are directly connected from those that are not, and also paths, in order to determine commonalities between nodes. There are many other patterns of interest that can be examined with this approach.

Finally, we consider the case where a graph is stored as an adjacency matrix, that is, as a table `Matrix(row, column, value)`, where the nodes of the graph have been numbered $1, \ldots, n$, and the entry `(i,j,v)` indicates that there is an edge between node `i` and node `j` with associated label or value `v`. Finding paths in the original graph can be achieved by multiplying the graph by itself, transposed. This is a simple operation in SQL, since multiplying matrices is straightforward and transposing the graph simply means using the row position as the column position and the column position as the row position.

First, assume we have two matrices $M1$ and $M2$ that can be multiplied (that is, the number of columns in $M1$ is the same as the number of rows in $M2$); then the product $M1 \times M2$ is simply

```
SELECT M1.row, M2.column, sum(M1.value*M2.value)
FROM M1, M2
WHERE M1.column = M2.row
GROUP BY M1.row, M2.column;
```

Exercise 4.20 Write an SQL query that produces the *sum* of matrices $M1$ and $M2$, assuming that they can be added (i.e. they have the same dimensions).

As noted, transposing is trivial:

```
SELECT  M1.column as row, M1.row as column, M1.value
FROM M1;
```

Multiplying matrix M by itself can be accomplished by using this schema with two copies of M and using the indices so that they represent transposition:

```
SELECT M1.row, M2.column, sum(M1.value*M2.value)
FROM M as M1,
     (SELECT M.column as row, M.row as column, M.value as value
      FROM M) as M2
WHERE M1.column = M2.row
GROUP BY M1.row, M2.column;
```

Exercise 4.21 Rewrite the query above to get rid of the subquery in the FROM clause. Hint: to do this, simply change how the indices are used.

Example: Boolean Matrices and Paths

Assume we want to find all paths of length up to k in a graph G and that G has been stored as the Boolean adjacency matrix M. We can compute M^k (M multiplied by itself k times), as follows.

```
WITH RECURSIVE PATHS(nodea int, nodeb int, steps int) AS
(SELECT a, b, 1 FROM MATRIX
 UNION
 SELECT a, b, steps + 1
 FROM MATRIX M, PATHS P
 WHERE M.b = P.nodea and steps < k)
SELECT *
FROM PATHS;
```

Note that if there are n nodes in G, we can compute all paths by using condition:

```
steps < (SELECT max(row) FROM MATRIX)
```

since `max(row)` (or `max(column)`, as this is a square matrix) is n, and (without loops) we cannot have any path longer than $n - 1$ steps.

4.7 Collaborative Filtering

Collaborative filtering is a family of techniques used by recommender systems. The basic idea is to filter information for an agent u using data about what other agents have seen/used/liked. A similarity distance between users establishes which users are similar to u; then their preferences are used to make recommendations for u. There are many variants of this idea.

Assume a table `Data(userid, itemid, rating)` with rows *(u,i,r)* if user u has given a rating r to item i. Then the simplest recommendation is: for a given user u and item i,

- item-item: find closest item i' to i, recommend i' to u.
- user-user: find closes user u' to u, recommend to u whatever u' likes.

To define 'closest' we need a distance. We introduced several distances in Sect. 4.3.1. Cosine is very popular among recommenders, so we use it in an example of item-item.

```
WITH similar_items(itemid, distance) AS
 SELECT D2.itemid, sum(D1.rating * D2.rating) /
               (sqrt(sum(D1.rating))*sqrt(sum(D2.ratings))
 FROM Data D1, Data D2
 WHERE D1.itemid = 'i' and D1.itemid <> D2.itemid
 GROUP BY D2.itemid
SELECT itemid
FROM similar_items
WHERE distance = (SELECT max(distance)
                  FROM similar_items);
```

This is the most similar item to item 'i.'

Exercise 4.22 Write an SQL query to determine the closest user to a given user u using cosine similarity over generic table `Data`.

Slope One is a family of algorithms for collaborative filtering [10] that uses the item-item approach. It uses an average of rating differences as distance, normalized by the number of common users. While this is a very simple measure (which makes it easy to implement and quite efficient), it sometimes performs on a par with more sophisticated approaches.

We again assume a table `Data(userid, itemid, rating)` and start by computing, for each pair of items, the average difference between their ratings as well as the number of common ratings. Using this, we can compute a recommendation for a user u on item i by applying the differences between i and other items and modifying u's rating on those other items accordingly.

```
WITH Diffs(itemid1, itemid2, freq, diff) AS
 (SELECT D1.itemid, D2.itemid, count(*),
        (sum(ud1.rating - ud2.rating))/count(*),
  FROM Data D1 join Data D2 on
        D1.userid = D2.userid and D1.itemid > D2.itemid
  GROUP BY D1.itemid, D2.itemid)
SELECT itemid2,  sum(freq) as freq,
    sum(freq*(diff + rating)) as pref,
    sum(freq*(diff + rating)) /sum(freq) as rating
FROM Diffs
WHERE itemid1 = 'i'
GROUP BY itemid2
ORDER BY rating
LIMIT 1;
```

This is the item closest to i according to Slope One.

Example: Slope One

We use the example from Wikipedia to illustrate the approach and predict Lucy's rating for item A. The example data (in a tidy format) is

Customer	Item	Rating
John	A	5
John	B	3
John	C	2
Mark	A	3
Mark	B	4
Lucy	B	2
Lucy	C	5

The table Diffs shows this result:

Item1	Item2	Count	Diff
A	B	2	0.5
A	C	1	3
B	C	2	−1

The 2 cases for A and B came from John (who gave A a 5 and B 3, for a difference of 2) and Mark (who gave A a 3 and B a 4 , for a difference of −1); both differences add up to 1, which divided by 2 cases yields a 0.5 average difference. Note that the symmetric entries (items A and B versus items B and A) are skipped by condition `itemid1 > itemid2`.

To predict Lucy's rate for item A, we use the difference between A and B (0.5) and add that to Lucy's rate for B (obtaining 2.5) and the difference between A and C (3) and add that to Lucy's rate for C (obtaining 8); these are weighted by the number of common ratings for A and B (2) and for A and C (1), to get $\frac{2(2.5)+1(8)}{2+1} = 4.33$, which is indeed the result of the query above over this data.

Exercise 4.23 Using the same data as the Wikipedia example, write a query to predict Mark's rating for item C.

Chapter 5
More SQL

In this chapter, we present some additional SQL operators in order to provide a well-rounded, complete overview of the language. We have left some operators for this chapter because they are not central to data analysis (set operations, WHERE clause subqueries) or because they are a bit more advanced (WINDOWS aggregates). At some points, we will revisit some solutions seen in previous chapters in order to show how to write them more efficiently (or, simply, in a different manner).

5.1 More on Joins

We already explained joins in Sect. 3.1.1. In this section, we add a bit more nuance to the behavior of this operator and present some versions of it that are useful in certain situations.

Joins can be combined with all other operations we have seen (including grouping, aggregation, and selection). These operations work on a single table, but that is exactly what the join produces, so one way to think of queries with joins is as follows: all tables mentioned in the FROM clause are joined; then, the query proceeds using this result exactly the way it did in the case of a single table.

There are two subtleties to understand about joins: the *existential* effect and the *multiplicative* effect. The existential effect refers to the fact that, when joining two tables, tuples in either table that do not meet the join condition are dropped from the result. As a consequence, a join may not have all the data in a table, and we should not assume otherwise.

Join: Existential Effect

Assume tables EMP(ssn,name,...) and SALES(essn,itemid,amount,date), where essn is a foreign key to EMP. The table SALES keeps track of the sales made,

A. Badia, *SQL for Data Science*, Data-Centric Systems and Applications,
https://doi.org/10.1007/978-3-030-57592-2_5

noting the employee who did the sales, the item sold, the amount, and the date. We wish to analyze employee's performance in the last quarter, so we issue this query:

```
SELECT ssn, name, sum(amount) as total
FROM EMP join SALES on (ssn = essn)
WHERE month(date) = 'November' and year(date) = '2015'
GROUP BY ssn, name
ORDER BY total asc
LIMIT 10;
```

The idea is to identify the bottom 10 employees by sales so we can look into their situations and see what led to this not-so-good performance. But we are missing something: suppose that some employees (because they were sick, or absent, or had a really bad month) did not sell anything at all. Will they show in the result with a 0 for `total`? They will not. Instead, we will see the bottom 10 employees *among those employees who had sales during that time period*. So employees are better off selling nothing than selling a small amount.

Why is this happening? Because the employees who sold nothing on that period do not make it past the join, as there are no matching tuples for them in the rows of SALES that pass the filter of the WHERE clause condition. This can be solved by using an *outer join*. On an outer join, we preserve tuples that do not match. We can preserve non-matching tuples in the first (left) table (a 'left outer join'), on the second (right) table (a 'right outer join'), or on both tables (a 'full outer join'). The syntax is

```
#Outer join syntax:
FROM table1 [LEFT|RIGHT] [OUTER] JOIN table2 ON condition
```

Non-matching tuples are missing some attributes, so those attributes are padded with NULLs. Thus, in our previous example, if we use

```
FROM EMP left outer join SALES on (ssn = essn)
```

we will end up with all employees; those that have sales will generate 'regular' tuples; those with no sales will be included in the answer in tuples where all attributes for SALES are padded with NULLs. As a result, the GROUP BY will generate a group for them with a single tuple, and `sum(amount)` will return a 0 (the only value to sum is a NULL).

This approach works especially well with optional attributes (see Sect. 2.1). When we want all the objects of a table together with all attributes (including attributes that may not have values for all objects), an outer join is in order. Remember, however, that non-matching tuples introduce nulls, so those must be cleared before any analysis.

Outer Join Use

Assume a table

```
VAERS(id, received-date, state, age, sex, symptoms, died,
      date-of-death, cause-of-death)
```

that keeps track of adverse effects to vaccinations.[1] Because the last two attributes only make sense when the attribute `died` is TRUE, we could separate this table into two, `VAERS(id, received-date,state,age,sex,symptoms)` and `DEATHS(id,date-of-death,cause-of-death)`, where the table `DEATHS` contains the ids only for those cases where a death occurred. This would avoid having to deal with nulls and 'irregular' data; however, we may want to do an analysis that requires all data combined (for instance, to determine mortality rates). If we simply join the tables

```
SELECT *
FROM VAERS JOIN DEATHS on VAERS.id = DEATHS.id;
```

we will miss all the cases that did not result in death; what we really want is a left outer join:

```
SELECT *
FROM VAERS LEFT OUTER JOIN DEATHS on VAERS.id = DEATHS.id;
```

Note that the attributes `date-of-death, cause-of-death` will have nulls for the cases where no death occurred, and that there is no attribute `dead` to mark such cases explicitly—although adding one is quite easy, as we have seen in past examples.

Another aspect of join that confuses newcomers to SQL is the *multiplicative effect*: when a tuple in table R matches *several tuples* in table S, this tuple gets repeated as many times as there are matches. This is a problem when we use aggregates that are *duplicate sensitive*.[2] In such cases, we may get extraneous results in our analysis.

Join: Multiplicative Effect

Assume the following situation: we manage a bank database, with tables

```
BRANCH(bid, name, address,...),
LOANS(lid,bid,amount,...),
ACCOUNTS(aid,bid,balance,...)
```

[1]Modeled after a real dataset available at https://vaers.hhs.gov/.

[2]An aggregate is *duplicate insensitive* when its result over a dataset does not change if duplicate values are removed; examples of this are MIN and MAX. An aggregate is *duplicate sensitive* when it is not duplicate insensitive; SUM, COUNT, and AVG are examples.

The first table lists the branches of the bank, with bid as its key. The second one lists each loan the bank currently has, with lid its key and bid a foreign key to BRANCH, indicating where the loan originated. The third table lists all accounts in the bank, with aid its key and bid a foreign key to BRANCH indicating where the account was opened. A bank manager is concerned that some branches are underperforming and asks us to find any branches where the total loan amount is larger than the total in deposits. We write the following SQL query:

```
SELECT bid, name
FROM BRANCH as B joins LOANS as L on (B.bid = L.bid)
     joins ACCOUNTS A on (B.bid = A.bid)
GROUP BY bid, name
HAVING sum(amount) > sum(balance);
```

Now imagine a situation where a branch has 2 loans and 3 accounts; for instance,

BRANCH		
Bid	**Name**	**Address**
1	Highlands	1500 Bardstown Road

LOANS		
Lid	**Bid**	**Amount**
10	1	1000
20	1	900

ACCOUNTS		
Aid	**Bid**	**Balance**
15	1	500
25	1	600
35	1	700

The join of all 3 tables is given by

Bid	Name	Address	Lid	Bid	Amount	Aid	Bid	Balance
1	Highlands	1500 Bardstown Road	10	1	1000	15	1	500
1	Highlands	1500 Bardstown Road	10	1	1000	25	1	600
1	Highlands	1500 Bardstown Road	10	1	1000	35	1	700
1	Highlands	1500 Bardstown Road	20	1	900	15	1	500
1	Highlands	1500 Bardstown Road	20	1	900	25	1	600
1	Highlands	1500 Bardstown Road	20	1	900	35	1	700

After the join, that branch appears 6 times; each loan in this branch is repeated 3 times and each account is repeated 2 times. As a consequence, neither sum (on amounts or balances) is correct.

Note that there is no other way to represent this information when all data is put together in a single table. What is happening here is that, while loans and accounts are (indirectly) related through the branch, they are orthogonal to each other. Thus, we cannot associate only some loans with some accounts; this could create a correlation between loans and accounts which does not truly exist. This is an example of data that cannot be really well analyzed in a single table.

What is the solution in this case? No extra operators are needed here; caution is. What we should do is compute each aggregate separately, one join at a time, in order to get the right results, which can then be compared. As usual, subqueries in the FROM clause are our friends:

```
SELECT bid, name
FROM (SELECT bid, name, sum(amounts) as loans
      FROM BRANCH as B joins LOANS as L on (B.bid = L.bid)
      GROUP BY bid, name) as TEMP1,
     (SELECT bid, name, sum(balance) as accounts
      FROM BRANCH as B joins ACCOUNTS A on (B.bid = A.bid)
      GROUP BY bid, name) as TEMP2
WHERE TEMP1.bid = TEMP2.bid and loans > accounts;
```

Note that one join multiplies the branch, since the relation between branches and loans is one-to-many (see Sect. 2.2), as is the relation between branches and accounts. This is ok; the repetition allows the GROUP BY to do its job by computing the aggregate value over each branch. What we need to avoid is mixing both one-to-many relationships together.

5.2 Complex Subqueries

The SQL standard allows for the expression of complex conditions in the WHERE clause that are specified using whole queries. Like the ones in the FROM clause, these 'embedded' queries are called subqueries. However, unlike the ones in the FROM clause, WHERE clause subqueries come in different flavors and with different predicates. We enumerate the types and give examples as follows:

- *aggregated subqueries*, which we have already seen, are subqueries with only aggregates in the SELECT clause. These are guaranteed to return a single value, which is compared against some attributes.

Example: Aggregated Subqueries

The query

```
SELECT name
FROM chicago-employees
WHERE  salary > (SELECT avg(salary)
                 FROM chicago-employees
                 WHERE salaried = True);
```

computes the average salary for all salaried employees; this result is then used to pick all employees that make more than this average.

- Subqueries with (NOT) IN: the IN predicate is used in combination with an attribute and a subquery. The subquery is expected to have a single attribute in its SELECT clause; as a result, the subquery returns a table with a single attribute on it—which we can think of as a list of values. The IN predicate checks whether the value of its first argument (attribute) in a given row is among those in the returned list of values.[3]

Example: IN Subquery

Assume tables
`EMPLOYEE(ssn,name,salary,job,dept)`
and
`DEPARTMENT(id,name,manager-ssn)`
and the query "list the names and salaries of all managers." The information about names and salaries is in table `EMPLOYEE`, but the information about who is a manager is in table `DEPARTMENT`, where `manager-ssn` is a foreign key to `EMPLOYEE.ssn`. We can write the following query:

```
SELECT name, salary
FROM EMPLOYEE
WHERE ssn IN (SELECT manager-ssn
              FROM DEPARTMENT);
```

The system runs the IN predicate on each row of `Employee`: it takes the value of `ssn` on the row, and it compares it to the list of ssn returned by the subquery, to see if it is one of them.

Naturally, NOT IN is simply the negation of IN: it returns true if the attribute value is not one of those in the returned list of values.

- Subqueries with (NOT) EXISTS: The EXISTS predicate takes a subquery and checks whether it returns an empty answer (no queries or not). EXISTS is satisfied if the subquery answer is *not* empty (i.e. if something exists in the answer).

Example: EXISTS Subquery

Assume the same database as the previous example, and suppose we want the name and salary of all employees in the Research (id "RE") department but only if there is a Marketing department (id "MK"); we are not sure whether this department exists. We can write the query

```
SELECT name, salary
FROM EMPLOYEE
```

[3]The SQL standard actually allows a more complex IN predicate, but most systems do not implement it.

```
WHERE dept = 'RE' and EXISTS (SELECT *
                              FROM DEPARTMENT
                              WHERE id = 'MK');
```

Note that the subquery uses '*', because the attributes returned are irrelevant; the EXISTS simply checks whether any rows are returned.

As before, NOT EXISTS is the negation of EXISTS; it is satisfied if the subquery returns an empty answer.

- Subqueries with ANY, ALL: the ANY and ALL predicates take an attribute and a subquery. The value of the attribute is compared to all values returned by the subquery; ANY requires that at least one comparison returns TRUE, while ALL requires that all comparisons return TRUE.

Example: Subqueries with ANY/ALL

Assume the EMPLOYEE-DEPARTMENT database. We want to find out which employees make more money than everyone in the Marketing department, so we run this query:

```
SELECT ssn, name
FROM EMPLOYEE
WHERE salary > ALL (SELECT salary
                    FROM EMPLOYEE
                    WHERE dept = "MK");
```

It is easy to see that several types of subqueries are redundant; for instance, the condition `attribute IN Subquery` is equivalent to `attribute = ANY Subquery`, and the condition `attribute NOT IN Subquery` is equivalent to `attribute <> ALL Subquery`. There are more equivalences, although some intuitive ones are disrupted by the presence of nulls.

Example: Subquery Equivalence and Nulls

The query used above to retrieve employees that make more than everyone in the Marketing department would seem to be equivalent to the query

```
SELECT ssn, name
FROM EMPLOYEE
WHERE salary > (SELECT max(salary)
                FROM EMPLOYEE
                WHERE dept = "MK");
```

but these queries may return different results when attribute `salary` in table `EMPLOYEE` contains nulls. The reason is that the comparison with ALL requires all comparisons between salaries return TRUE, but the result is unknown for nulls; this causes the ALL predicate to fail. However, the aggregate `max` will happily compute a result while ignoring any nulls. As far as there are also some non-null values in `salary`, this result will be non-null and will be used by the comparison. Thus, the second query may or may not return something, depending on the data in table `EMPLOYEE`.

Subqueries in the WHERE clause can be *correlated*. Such subqueries mention an attribute that comes from the outer query that uses them. For instance, assume that we want to find all employees who make above their department's average salary; one way to write this is

```
SELECT name
FROM Employee E1
WHERE salary > (SELECT avg(salary)
                FROM Employee E2
                WHERE E2.dept = E1.dept);
```

This query is understood as follows: results from E1 (the copy of Employee used in the outer query). On each row, we return employee x's name if x's salary is greater than the average salary, calculated over the rows of E2 (another copy of Employee) that match x's department (i.e. the average salary in x's department). All types of WHERE clause subqueries can be correlated. In fact, it is very common for some of them (like EXISTS) to be used primarily in correlated contexts.

Thanks in part to the subquery predicates, there are several different ways to write most queries in SQL. The analyst has a choice as to how and when to use subqueries, as we can see with some simple examples.

Example: Moving Subqueries from FROM to WHERE

Example 3.1.2 of Sect. 3.1.2 is repeated here with a subquery in the WHERE clause, instead of in the FROM clause.

```
SELECT name
FROM chicago-employees
WHERE  salary > (SELECT avg(salary) as avgsal
                 FROM chicago-employees
                 WHERE salaried = True);
```

The system will first evaluate the subquery, obtaining a value for the average (mean) of across all values of table `Chicago-employees`. It will then use this value to evaluate the condition. Suppose, for instance, that the average found was 45.6; the condition in `WHERE` will be treated as `salary > 45.6`. Note that the same table is mentioned in both `FROM` clauses; this is not unusual at all. One can think of this as having two copies of the same table, being used independently to evaluate the subquery and the main query.

Most subqueries in SQL can be avoided. Aggregated, non-correlated subqueries can be used in the WHERE clause but they can also be put in the FROM clause and its result is used. For instance, the query above was written originally with a subquery in the FROM clause.

Aggregated and correlated subqueries can also be moved to the FROM clause, but we need to add a grouping to simulate the effects of the correlation. For instance, the query above can be rewritten as

```
SELECT name
FROM Employee, (SELECT dept, avg(salary) as avgsal
                FROM Employee
                GROUP BY dept)
WHERE salary > avgsal;
```

Queries with IN and EXIST can be turned into joins, whether they are correlated or not. For instance, the example with IN above can be also written as

```
SELECT name, salary
FROM EMPLOYEE, DEPARTMENT
WHERE ssn = manager-ssn;
```

Also, queries with NOT IN and NOT EXIST can be rewritten using EXCEPT, as will be explained when set predicates are introduced in Sect. 5.4.

Of note, queries with EXISTS and NOT EXISTS can be turned into aggregated subqueries, whether they are correlated or not. For instance, the example with EXISTS above can be written as

```
SELECT name, salary
FROM EMPLOYEE
WHERE dept = 'RE' and 0 < (SELECT count(*)
                           FROM DEPARTMENT
                           WHERE id = 'MK');
```

We change the SELECT clause in the subquery from '*' to `count(*)`, which returns the number of rows in the answer. If this number is greater than 0, then there is at least one row in the answer, so EXISTS is satisfied (using equality instead of < will express NOT EXISTS).

In the end, there is no real need to use subqueries in the WHERE clause in SQL. They were introduced early in the standard and the desire to keep the standard backward compatible has resulted in subqueries still being there even though they are not strictly necessary.

5.3 Windows and Window Aggregates

Windows have been added to the SQL standard to give more flexibility than GROUP BY allows. A *window* is a set of rows from a table, specified by the user, on which certain calculations are performed. In a sense, they are similar to groups, since a table can be partitioned into a set of windows. However, unlike groups, windows are not collapsed to a row; it is possible to operate *within* them, as the individual rows that make up a window can be accessed and manipulated as convenient.

Window functions can only appear in SELECT and ORDER BY clauses. To specify a window, we use a WINDOW clause, which has the syntax

```
WINDOW name AS (PARTITION-BY ...ORDER-BY ...FRAME ...)
```

These three components, which are all optional, are understood as follows:

- PARTITION BY takes a list of attributes as argument and is equivalent to GROUP BY: it creates the windows by putting together all rows of the input table that have the same values for the specified attributes.
- ORDER BY takes a list of attributes as arguments and is equivalent to an ORDER BY clause in that it sorts the tuples within a window. Each window is sorted separately, so there is a first row, second row, etc. on each window. The user can also specify where to put nulls (first or last).
- FRAME is an expression that specifies a subset of rows within each window to which an aggregate applies. That is, in each window, the aggregate takes as input only the rows denoted by the FRAME, not all of them. A common way to denote certain rows is to use the previously defined order. It is assumed that, as the system applies an aggregate over a window, it will scan it row-by-row, so that it will advance from the first row to the last ('first' and 'last' defined according to the order). As it does that, there is a 'current' row that is used as a reference. The frame specifies which rows around the current row are involved in computing the aggregate—that is, it defines a neighborhood of the current row. This is defined by ROWS (number of rows before/after current one) or RANGE (values in ordering attribute are in a range relative to current one); such rows can be PRECEDING or FOLLOWING the current row.

A few examples will show the basic idea:[4] assume a table

```
Sales(storeid, productid, day, month, year, amount)
```

The query

```
SELECT storeid, month, amount, avg(amount) over w
FROM Sales
WHERE month BETWEEN 2015/09 and 2015/12
WINDOW w AS (PARTITION BY storeid, ORDER BY month,
             ROWS 2 PRECEDING);
```

computes the *moving average* over 3 months. Here the FRAME is

```
ROWS 2 PRECEDING
```

which means 'the current row and the two rows before it (in the order given, i.e. by month).' If the FRAME used were

```
ROWS BETWEEN 1 PRECEDING AND 1 FOLLOWING
```

[4]These examples come from the description of the SQL standard on SIGMOD Record.

this would also give a moving average, but this time using the past, present ('current'), and next month. Finally, the window

```
RANGE BETWEEN INTERVAL '1' month PRECEDING
          AND INTERVAL '1' month FOLLOWING)
```

we would get the same as in the previous case: the frame consists of the current row, the previous one (`'1' month PRECEDING`), and the next one (`'1' month FOLLOWING`). There is a subtle difference, though:

- ROWS is physical aggregation; if there are gaps or repetitions on the data, it will still pick the previous, current, and following rows, giving dubious results.
- RANGE is logical aggregation: it will skip gaps or repetitions and always find the values that precede and follow the current month. However, RANGE can only be used with one numerical grouping aggregate.

The keyword UNBOUNDED can be used in ROWS instead of a number; it means to use all rows up to the window boundary. For instance, the FRAME

```
ROWS BETWEEN UNBOUNDED PRECEDING AND CURRENT ROW
```

would give a *cumulative* aggregate, as each aggregate would include all rows from the beginning (relative to ORDER) of the group until the current one, which progresses row-by-row from beginning to end (again, relative to ORDER).

Multiple windows can be defined in the same query, in the same WINDOW clause:

```
SELECT storeid, productid,
       sum(amount) OVER everything,
       sum(amount) OVER bystore,
       sum(amount) over byproduct
FROM sales
WINDOW everything AS (),
       bystore AS (PARTITION BY storeid),
       byproduct AS (PARTITION BY productid);
```

This query computes three sums: one over the whole table, one per store (partitioned by `storeid`), and one per product (partitioned by `productid`). Note that this can be written with just regular grouping but would require three SELECT statements.

Exercise 5.1 Write the equivalent to the query above using regular grouping. Hint: use three subqueries in WITH or FROM.

One window can also be defined in terms of another window:

```
SELECT storeid, month, sum(amount) over W2a,
                       avg(amount) over W2b
FROM sales
WHERE month BETWEEN 2001/09 and 2001/12
WINDOW w AS (PARTITION BY storeid, ORDER BY month),
```

```
        w2a AS (w ROWS BETWEEN UNBOUNDED PRECEDING
                           AND CURRENT ROW),
        w2b AS (w ROWS BETWEEN 1 PRECEDING AND 1 FOLLOWING);
```

This query gives the cumulative sales per month and a centered average per month. Windows `w2a` and `w2b` are based on window `w`, so they 'inherit' their PARTITION and ORDER from `w`.

Exercise 5.2 Write a query over `sales` that computes the moving average sale amount over 5 months per customer.

The use of windows on analytics relies on the fact that it provides us with this fine grained control over how to compute aggregates (so we can get *cumulative* and *moving* aggregates) but also on the fact that they can be used as part of a query where arbitrary attributes (whether mentioned in the window or not) can be retrieved. This makes it more flexible then GROUP BY. For instance, our first example of this section retrieved both the individual sale amount and a running average computed for each store. This means that the result will display, for each store, as many rows as sales there were in that store, with the individual amount of each, and the computed aggregate added to each row. To do this with GROUP BY we need to use a subquery, due to the restrictions on GROUP BY syntax.

We now show how window aggregates can be used to accomplish some of the tasks already explained in the book, but expressed in a more concise (and sometimes more efficient) manner. As a simple example, in Sect. 3.2 we calculated cumulative totals as follows:

```
SELECT V2.Value, Sum(V.Value) as Cumulative
FROM Values as V2, Values as V
WHERE V.Order <= V2.Order
GROUP BY V2.Value;
```

With windows, this can be accomplished without a Cartesian product, which is more efficient:

```
SELECT V.Value,
       sum(Value) OVER(PARTITION BY Order
                       ORDER BY Order
                       ROWS BETWEEN UNBOUNDED PRECEDING
                            AND CURRENT ROW) AS Cumulative
FROM Values as V
GROUP BY V.Value;
```

Rankings come in handy for computing ranking-based correlation measures like Spearman's or Kendall's (see Sect. 3.2.2). The window function `RANK()` creates an explicit attribute with the ranking of a row based on the ordering specified in the ORDER BY:

```
SELECT *,
       rank() OVER (PARTITION BY store-id
                    ORDER BY amount DESC)  AS position;
```

will return a table with an attribute called position with values 1, 2, ... on each group created by store-id, with the value in position based on the amount attribute: the largest amount is 1, the second largest is 2, ... (note that we are sorting in descending order). The ranking is restarted at 1 for each group (each store).

Using this, we can simplify the computation of top k results, which we did earlier using ORDER BY and LIMIT:

```
SELECT *,
       rank() OVER (PARTITION BY store-id
                    ORDER BY amount DESC)  AS position
WHERE position < 11;
```

will retrieve, for each store, the top 10 sales by amount (i.e. the 10 largest sales). However, it must be noted that this query may return more than the intended 10 rows in case of ties. To deal with this, SQL distinguishes between RANK() and DENSE_RANK(), which does not skip ranks. As a simple example, if in some stores the top 4 sales are 1000, 900, 900, and 800, RANK() will produce 1, 2, 2, 4 and DENSE_RANK() will produce 1, 2, 2, 3.

Explicit ranking could be used for calculating percentiles. However, many systems (including Postgres) have 'ordered aggregates' that do this directly:

```
percentile_disc(fraction) WITHIN GROUP (ORDER BY column(s))
```

This is the *discrete percentile* function; it returns the first input value whose position in the ordering equals or exceeds the specified fraction (fraction must be a value between 0 and 1.0). There is also

```
percentile_cont(fraction) WITHIN GROUP (ORDER BY column(s))
```

This is the *continuous percentile* function; it returns the value corresponding to the specified fraction in the ordering. Note that percentile_disc returns a value from the dataset, while percentile_cont may create a value by interpolation if needed. Hence, we can use

```
SELECT percentile_cont(0.5) within group (order by A)
FROM Dataset;
```

to get the median value of column A. We can also use this as an aggregate function, in a query with GROUP BY,[5] to find medians within groups:

```
SELECT storeid,
       percentile_cont(0.5) WITHIN GROUP (ORDER BY amount)
                            AS StoreMedian
FROM sales
GROUP BY storeid;
```

Exercise 5.3 Write a query over sales that computes both the mean and the median sale amount per customer.

---

[5]The percentile aggregates are not windows aggregates, so they cannot be used with OVER.

A small aside: even though computing the mode is not hard, Postgres also has a window aggregate for that:
`mode() WITHIN GROUP (ORDER BY attrib)`
will compute the mode of attribute `attrib`.

Checking outliers with MAD is doable in Postgres by calculating the median with `percentile_cont`. Assume that in the people dataset we have a `height` column that contains some suspicious values. We can use

```
SELECT percentile_cont(0.5) WITHIN GROUP
                            (ORDER BY height - median)
FROM Dataset,
     (SELECT percentile_cont(0.5) WITHIN GROUP
                                  (ORDER BY height)
             as median
      FROM Dataset);
```

to get the MAD value and decide if those values are outliers or errors.

To find quartiles with this approach, we could use

```
SELECT storeid,
       percentile_cont(0.25) WITHIN GROUP (ORDER BY amount)
  AS quart1,
       percentile_cont(0.5) WITHIN GROUP (ORDER BY amount)
  AS quart2,
       percentile_cont(0.75) WITHIN GROUP (ORDER BY amount)
  AS quart3,
       percentile_cont(1.0) WITHIN GROUP (ORDER BY amount)
  AS quart4
FROM sales
GROUP BY storeid;
```

It is possible to also use the `ntile(`n`)` function, which sorts the values and breaks them into n buckets. To get quartiles with this approach, it is enough to use

```
SELECT DISTINCT storeid,
       ntile(4) OVER (PARTITION BY storeid ORDER BY amount)
                          as quartile
FROM sales;
```

The function creates an attribute called `quartile` with values 1, 2, 3, 4. A value i is assigned to each tuple, depending on whether the amount is on the i-th quartile, within each store group. Once this is done, the result can be used to get an aggregate within each quartile; for instance, the largest/last value would be close to the result from using `percentile_disc` as above, with values at each .25 of the total.

Exercise 5.4 Calculate deciles using `percentile_disc` or `percentile_cont`.

Exercise 5.5 Calculate deciles using `ntile`.

Another example of the usefulness of this additional aggregates is the computation of the *k (k%) trimmed mean*. Recall that this is the mean calculated excluding the top k (or $k\%$ and the bottom k (or $k\%$) of all values, in order to defuse the influence of outliers. Support, for instance, that we want to exclude the top and bottom 5%; this can be achieved with

```
SELECT avg(attr)
FROM Data,
  (SELECT percentile_disc(.05) WITHIN GROUP (ORDER BY attr)
          as mincut,
          percentile_disc(.95) WITHIN GROUP (ORDER BY attrb)
          as maxcut
   FROM Data) as cuts
WHERE attr > mincut and attr < maxcut;
```

If we want to exclude the top (and bottom) k values, we can use `rank()` to sort them and then use the new attribute created by `rank()` to exclude the appropriate number of values. Note that excluding the top k is easy with descending order, but the bottom k requires some additional computation to determine which values to exclude (of course, the situation using ascending order is symmetrical).

Exercise 5.6 Write the query to calculate the top 10 trimmed mean on the generic `Data` table.

Windows aggregates can be used to compute distances more efficiently too. We saw in Sect. 4.3.1 that computing distances between any two points in a dataset required an expensive Cartesian product, but we can avoid that in some cases. Assume in table `Data` there is a numerical attribute `A` and we want to compute the distances between any two points using their differences in value of `A`; instead of writing

```
SELECT *, abs(R.A - S.A)
FROM Data as R, Data as S;
```

we can use window aggregate `lag`, which computes the difference between a row and its preceding one (where 'preceding' is determined, as usual in windows aggregates, by some ordering). This is accomplished with

```
SELECT *, lag(A) OVER (PARTITION BY id, ORDER BY id, A)
          as previous,
          A - lag(A) OVER (PARTITION BY id, ORDER BY id, A)
          as diff
FROM Data;
```

The attribute `diff` can be used as a distance based on attribute `A`. This can be extended to distances based on several attributes by computing (and combining) separate lags for each attribute. Of course, distances more sophisticated than `abs(R.A - S.A)` may require additional computation.

Exercise 5.7 Repeat the clustering algorithm based on random pivots of Sect. 4.3.1 using `lag` to calculate distances.

Exercise 5.8 Repeat the clustering algorithms based on thresholds of Sect. 4.3.1 using `lag` to calculate distances.

Finally, we introduce a couple of extensions of regular GROUP BY, ROLLUP, and CUBE. Even though not really window functions, these may come in handy in some cases. A query using
GROUP BY ROLLUP($attribute_1, \ldots, attribute_n$)
is equivalent to a query using all these groupings:

- no grouping at all;
- a grouping with $attribute_1$;
- a grouping with $attribute_1$, $attribute_2$;
- ...
- a grouping with $attribute_1$, $attribute_2$, ..., $attribute_{n-1}$; and
- a grouping with $attribute_1$, ..., $attribute_n$.

That is, one grouping per *prefix* of $attribute_1$, ..., $attribute_n$.
Conversely, a query using
GROUP BY CUBE($attribute_1, \ldots, attribute_n$)
is equivalent to a query using all groupings that can be done with any subset of $attribute_1$, ..., $attribute_n$. A simple example will illustrate their usage.

Example: Rollup and Cube

Recall table SALES(storeid, productid, time, day, month year, amount).

- the query

```
SELECT year, month, day, sum(amount) as total
FROM Sales
GROUP BY ROLLUP(year, month, day);
```

will calculate aggregates over the following groups : (year, month, day) (sums per day), (year, month) (sums per month), (year) (sums per year), the sum of the whole table. The result table will look as follows:

Year	Month	Day	Total
Some year	Some month	Some day	Total for year, month, day
		...	
Some year	Some month	Null	Total for year and month
		...	
Some year	Null	Null	Total for year
		...	
Null	Null	Null	Total for whole table

Since each one of those is a different grouping, this query is equivalent to 4 queries with regular GROUP BY (in general, a query with roll-up on n attributes is equivalent to $n + 1$ queries with regular GROUP BY).

- the query

```
SELECT year, month, day, sum(amount)
FROM Sales
GROUP BY CUBE(year, month, day);
```

will form the following groups: `(year, month, day)`, `(year, month)`, `(year, day)`, `(year)`, `(month, day)`, `(month)`, `(day)`, `()`. Again, each one of those requires a separate group by, so the query with CUBE is equivalent to 8 queries with regular grouping(in general, a query with CUBE on n attributes is equivalent to 2^n queries with regular GROUP BY). The result table will look as follows:

Year	Month	Day	Total
Some year	Some month	Some day	Total for year, month, day
		...	
Some year	Some month	Null	Total for year and month
		...	
Some year	Null	Null	Total for year
		...	
Null	Null	Null	Total for whole table
Some year	Null	Some day	Total for year and day
		...	
Null	Some month	Some day	Total for month and day
		...	
Null	Some month	null	Total per month
		...	
Null	Null	Day	Total per day

While queries with ROLLUP and CUBE can be written without them, they clearly make calculations simpler to write (and, in most cases, much more efficient to compute).

Exercise 5.9 Simulate the rollup query in the example without using ROLLUP, with simple GROUP BY. Hint: use the union of several queries (see next section), one for each specific grouping.

Exercise 5.10 Simulate the cube query in the example without using CUBE, with simple GROUP BY. Hint: use the union of several queries (see next section), one for each specific grouping.

Exercise 5.11 Assume the spreadsheet

State/Year	2000	2001	...	
AL	500	350	...	AL-total
AK	500	350	...	AL-total
...	...	...	...	...
	2000-total	2001-total	...	Total

We know that this would be represented as a table with schema (`state,year, total`) and that all values with 'total' in the name are not raw data—so they would not be part of the table. Calculating those *margin* totals is not complicated, but it used to require 3 separate queries (one for the row totals, one for the column totals, one for the whole total). Write a single query that will compute all totals at once with one of our new found friends.

5.4 Set Operations

SQL allows *set operations*; in particular, taking the union, intersection, and difference of two tables is allowed. This is expressed in SQL by using two queries and combining them with the keywords UNION, INTERSECT, EXCEPT. For instance, to take the union of two queries we write

```
SELECT ... FROM ... WHERE ...
UNION
SELECT ... FROM ... WHERE ...
```

The system will run the first (topmost) query, as well as the second (bottom) one; it will then combine the rows from both answers into a single table. If we use INTERSECT instead of UNION, the system computes the intersection of both answers (i.e. the rows that are present in both answers). If we use EXCEPT instead, we get all the rows of the first (topmost) answer that are *not* present in the second (bottom) answer.

However, the system requires that the tables to be combined are *schema compatible*. Two tables are schema compatible iff they have the same number and type of attributes. For instance, if the answer to a table includes three attributes, one of them being an integer, the second a string, and the third a date, a schema compatible table will also have three attributes, and the first one will be an integer, the second a string, and the third one a date. This ensures that the tables can be meaningfully combined into one. Note that this refers to the tables that are input to the UNION (INTERSECT, EXCEPT) operators, that is, to what is used in the SELECT clause of the queries, not to the tables in the FROM clause of the queries.

Example: UNION Operation

Assume we have several tables named
`PsychologyRank(school-name, state, type, ranking-position)`
for schools ranked according to their Psychology programs; another table
`EconomyRank(school-name, state, type, ranking-position)`
where the schools are ranked according to their Economy programs; another table
`HistoryRank(school-name, state, type, ranking-position)`
and so on.

```
SELECT 'Psychology' as program, *
FROM PsychologyRank
UNION
SELECT 'Economy' as program, *
FROM EconomyRank
UNION
SELECT 'History' as program, *
FROM HistoryRank;
```

Note that we have applied two UNIONs to three tables; as their set counterparts, the operations are binary (take two arguments) but yield an object of the kind (table or set) and so can be used several times. Moreover, the union is associative and symmetric, so the order of tables does not matter (intersection is also associative and symmetric, but the difference is neither). Note also that, if the schemas of the tables being analyzed are not exactly the same, or we do not want all the data, we can use the SELECT clause to pick the common (or desired) attributes of similar tables.

Exercise 5.12 Assume we have data about New York City real estate sales, and we have 5 different datasets, one for each one of the 5 boroughs that are part of the city: Manhattan, Brooklyn, Queens, The Bronx, and Staten Island. Each dataset is in a table with the name of the borough and a similar schema: `(address, type, date-sold, amount-sold)`. Put all data together in a single table with schema `(borough,address,date-sold,amount-sold)`.

Example: INTERSECT and EXCEPT Operations

Assume the same tables as in the previous example and suppose that this time we just want to know if there are schools (names) that are tops on both Psychology and Economy. The following query will take care of this:

```
SELECT name
FROM PsychologyRank
INTERSECT
SELECT name
FROM EconomyRank;
```

If we want schools that are top ranked in Psychology but not in Economy, we use this instead:

```
SELECT name
FROM PsychologyRank
EXCEPT
SELECT name
FROM EconomyRank;
```

Note how this time we focused on attribute name; this made the results schema compatible while giving us all the information required.

As stated earlier in Sect. 5.2, several complex predicates express negation, including NOT IN and NOT EXISTS. These conditions can also be written using EXCEPT instead. The idea is to write a query Q1 EXCEPT Q2, where Q2 is the subquery of the NOT IN (or NOT EXISTS) condition, and Q1 is the main query. Thus,

```
SELECT attrA
FROM Table1
WHERE attrB NOT IN (SELECT attrC
                    FROM Table2);
```

becomes

```
SELECT attrA
FROM (SELECT attrB
      FROM Table1
      EXCEPT
      SELECT attrC
      FROM Table2) as Temp, Table1
WHERE Temp.attrB = Table1.attrB;
```

When attrA = attrB, the transformation is even simpler, as the next example shows.

Example: NOT IN Transformed into EXCEPT

Assume we are running several trial experiments over a population, and table Participation(subject-id, trial-id, date) which keeps track of which subjects participate in which trials. We want to get all subject that participated in trial 'ACK' but not in trial 'PYL.' We can write this using NOT IN as

```
SELECT subject-id
FROM Participation
WHERE trial-id = 'ACK' and
      subject-id NOT IN (SELECT subject-id
                         FROM Participation
                         WHERE trial-id = 'PYL');
```

but we can also write it with EXCEPT as

```
SELECT subject-id
FROM Participation
WHERE trial-id = 'ACK'
EXCEPT
SELECT subject-id
FROM Participation
WHERE trial-id = 'PYL';
```

5.5 Expressing Domain Knowledge

Domain knowledge (sometimes called *subject matter knowledge*) refers to the facts and constraints that we have about whatever real-world domain the data refers to. Such knowledge usually allows us to predict in advance what kinds of attributes to expect and, for each attribute, what values are normal or common. For instance, in a medical database of patients, we can expect attributes expressing measurements of medically significant factors like blood pressure, cholesterol levels, etc. A *subject matter expert* (in this case, an MD) could tell us what are typical and atypical values for many such attributes. This is very useful for EDA, since it makes it much easier to determine if there are errors in our data, or if we have outliers or missing values.

Another way in which this domain knowledge can be used is by telling the system to check attribute values for us. For instance, for closed domain values, we can tell that only certain values are acceptable. In this case, other values should be rejected. For this situation we have the CHECK statement. This statement can be used when creating a table or can be added later to the table definition using ALTER TABLE. It has the syntax

```
CHECK condition
```

where `condition` is a condition similar to those of a WHERE clause. In most systems, referencing other tables is not allowed; the condition is restricted to the table that contains the CHECK. Even with this limitation, CHECKS can be extremely useful to find 'bad' data.

Example: CHECK Statement

Assume, for the New York flights database, that we know (since this is a dataset of flights *from* New York) that the only valid values for attribute `origin` are 'EWR,' 'LGA,' and 'JFK.' We also know that the arrival time should be later than the departure time, attribute `month` should be a number between 1 and 12, and `day` should be a number between 1 and 31.[6] In general, constraints that involve only

[6]Note that this is still not enough to catch bad dates; transforming these values into dates (see Sect. 3.3.1.3) is the right thing to do.

one attribute value, or several attribute values in the same row, are expressible with CHECK statements.

```
CREATE TABLE NY-FLIGHTS(
flightid int,
year int,
month int CHECK (month BETWEEN 1 and 12),
day int CHECK (day BETWEEN 1 and 31),
dep_time int,
sched_dep_time int,
dep_delay int,
arr_time int,
sched_arr_time int,
arr_delay int,
carrier char(2),
flight char(4),
tailnum char(6),
origin char(3) CHECK (origin IN ('EWR', 'LGA', 'JFK')),
....
CHECK (sched_dep_time < sched_arr_time));
```

Note that the check on departure and arrival time is added at the end of the attribute list, while other checks that involve a single attribute are added after the attribute definition. Note also that we have used < to express 'earlier than' because both the departure and arrival time are expressed as integers.

Exercise 5.13 In the `NY-Flights` data, there are several constraints that should hold of the data in each flight (each record):

- `arr_time` should also always be later than `dep_time` and `sched_dep_time`.
- `dep_delay` should be the difference between `dep_time` and `sched_dep_time`.
- `arr_delay` should be the difference between `arr_time` and `sched_arr_time`.

Write a more complete `CREATE TABLE` statement that enforces all these constraints using CHECKs.

In most systems, there exists the option of *disabling* checking when loading data. The main advantage of this is speed: if there are CHECK statements in a table, the system tests them every time there is an insertion into the table; when loading data from a file, this means one check per row (see Sect. 2.4.1). This slows down the loading; that is why many systems provide a way to circumvent the checking. The price to pay for this is that we may upload bad data into our database and we will have to search for it manually: if we enable the CHECKS back once the data is loaded, the system will not test the data already in the table. Thus, unless we are loading a very large amount of data and have limited time, it is a good idea to let the system check the data for us as it loads the table. We will still have to deal with the errors manually, but at least we know exactly which data causes problems.

Chapter 6
Databases and Other Tools

In this chapter, we show how to connect to a relational database from R and from Python. Unlike the rest of the book, this chapter assumes that the reader is already familiar with R and/or Python.

Communication between R and Python and a database is very formulaic; once the basic pattern is understood, most of the work is to figure out what we want to extract from the database. In the following examples, we stick with very simplistic interactions, in order to make the basic pattern clear; we use mostly examples from the documentation (in `cran.r-project.org` or the package documentation). The important point to remember is that all the interactions with the relational database are carried out in SQL; therefore, everything shown in the rest of the book can be applied here.

6.1 SQL and R

There are many ways to work with databases within R. Among them,

- Using the DBI library.
- Using packages dplyr and dbplyr.
- Using package sqldf.

In this section we describe each one of them. We assume that the reader is familiar with the basics of R; in particular, knowledge of *data frames* and their manipulation is assumed. We show how to translate much of the SQL we have seen in the book into R, but we will not go into full detail.

In our examples, we use the `mtcars`, `warpbreaks`, and `iris` datasets, which come with the standard R distribution,[1] as well as the New York flights dataset we

[1] Use `data()` to list the data frames available (built-in) in R.

A. Badia, *SQL for Data Science*, Data-Centric Systems and Applications,
https://doi.org/10.1007/978-3-030-57592-2_6

have been using all along, so that the similarities between SQL and R are made clear. We show R commands as they should be typed into the system; comments are in lines started with '#'. Occasionally we show the answer of the systems; such lines are prefixed with '>.'

6.1.1 DBI

DBI is a package that provides basic interfacing with databases. As such, it is used by some of the other interfaces, as will become apparent. Here we show only some basic DBI functionality.

DBI connects with a relational database and interacts with it using SQL. It uses packages called `drivers` to handle interaction with a particular database (Postgres, MySQL, etc.). This allows DBI to hide a large amount of low-level details and make the interaction with the database much simpler than it would otherwise be.

The package provides a collection of functions to support each step in the interaction with the database. These steps are:

1. connection: to establish a connection with the database, the function `dbConnect` takes as arguments a `driver` (depending on the type of database one is connecting to), a `host` (addresses where to find the database), `dbname` (database name), `user` (username), and `password` (ditto). This function returns an object of type 'connection,' which is used as a parameter by other functions that make use of this connection.
2. access: to access the database, the user has at her disposal a series of functions that take, as one of the parameters, the connection object (this makes sure the right database is accessed) and as another one a string representing an SQL command, which tells the database what we want to do. Depending on the required access, the most common functions are:

   - for metadata, `dbListTables` takes a connection object as parameter and lists all the tables available in the database accessed; `dbListFields` takes a connection object and a table name as parameters and returns the schema of the table. There is also metadata associated with a result returned by a query (see below); function `dbColumnInfo` takes as argument a result set and gives information about its schema.
   - to send data from R to the database, function `dbWriteTables(connection, table-name, data-frame)` will create a table in the connected database with the name given by the second argument and will copy to it the data in the third argument. The schema of the table is deduced from the data frame.
   - to get data from the database into R, function `dbReadTables(connection, table-name)` returns a data frame containing the data in the table named by the second parameter.
   - to make changes in the database, functions `dbCreateTable`, `dbRemoveTable` do exactly what the name suggests, creating and dropping tables. The first

one takes as arguments a connection, a name for the table, and (optionally) a named vector describing the attributes of the table (the name is the name of the attribute, the value is its data type), or a data frame. When using a data frame, a schema is inferred from the data in the data frame. Also, function `dbExecute` takes as arguments a connection and a string representing a "modify" SQL statement, that is, INSERT, DELETE, or UPDATE.

- to retrieve some data selectively from the database, we use `dbSendQuery`, which takes as parameters a connection and a string with an SQL SELECT statement. This function returns an object called a *result set*, which can then be used to access whatever data was returned by the query. The data in the result set is typically accessed by *iterating* (AKA *looping*) through it, that is, accessing each record one by one. The iteration is accomplished by combining several functions:
  - function `dbFetch`, which takes as arguments a result set and (optionally) a number n. The function returns records from the result set, which can be assigned to a data frame in R. If $n > 0$, it returns n records from the result set (assuming there are at least n records; if there are fewer than n, it just returns whatever records are available). If $n < 0$, all remaining records are returned. If n is not used, all records in the result set are returned.
  - function `dbHasCompleted` returns True if fetching has exhausted all the rows in the return set; False if there are any still left.

  A typical iteration is shown in the example below.

3. Finally, `dbDisconnect` takes as parameter the connection object and 'finishes' it, closing the connection.

Example: Connecting to a Database with DBI

Typical code used in a database connection with `DBI`:

```
library(DBI)
# Connect to a MariaDB remote database
con <- dbConnect(RMariaDB::MariaDB(), host = "hostname.com",
                 user = "username", password = "password")

# Connect to a MariaDB database local database
con <- dbConnect(RMariaDB::MariaDB(), dbname = "mydb")

# Create and connect to an in-memory RSQLite database
con <- dbConnect(RSQLite::SQLite(), dbname = ":memory:")

#find out what tables exist in the database, if any
dbListTables(con)

#write data frame mtcars to database table "mtcars"
dbWriteTable(con, "mtcars", mtcars)
```

```
#find out attributes of table
dbListFields(con, "mtcars")

# read table "mtcars" into R data frame
df = dbReadTable(con, "mtcars")

# You can fetch all results at once:
res <- dbSendQuery(con, "SELECT * FROM mtcars WHERE cyl = 4")
dbFetch(res)

# Or a chunk at a time
res <- dbSendQuery(con, "SELECT * FROM mtcars WHERE cyl = 4")
while(!dbHasCompleted(res)){
  chunk <- dbFetch(res, n = 5)
  print(nrow(chunk))
}

#disposes of the results after iteration
dbClearResult(res)

dbDisconnect(con)
```

The loop shown above is the typical way to examine the result returned by a query, using the parameter n to control how many records we retrieve at a time. This is especially useful for very large results, since one of the limitations of R is that it does not do well when it cannot have all data in a data frame (or any other structure) in memory. A similar approach is used in other interfaces, as we will see.

The DBI package contains many other functions that allow the user to 'remotely control' a database, that is, to connect to it and send all sorts of SQL statements, achieving the same effects as if the user were directly connected to the database. In addition, the package makes very easy to transfer data between R and the database. Note, though, that all the interactions with the database are handled using SQL.

DBI (and several other packages) uses (sometimes by default) a database called SQLite. This system, unlike Postgres and MySQL (and most any other database system) does not require any setup and can be used from within other applications without any configuration. This makes it a very popular database for R and other systems to call.

Example: Using SQLite

The approach is similar to using other databases with DBI:

```
install.packages("RSQLite")
library(RSQLite)

#create a new, empty database using the given file
conn <- dbConnect(SQLite(),'mycars.db')
```

```
#create a table with schema deduced from data
dbWriteTable(conn, "cars", mtcars)

#one can also create tables the traditional way
dbGetQuery(conn, 'CREATE TABLE test_table(id int, name text)')
```

SQLite stores all data in a single file and does not enforce data types or foreign keys, so it is possible to store 'wrong' data on it. Its use is mainly for lightweight, exploratory analysis. Most standard SQL querying is available:

```
dbGetQuery(conn, "SELECT * FROM cars WHERE mpg > 20")
```

If one already has a Postgres server up and running, using it instead is very simple:

```
#installing the library
install.packages("RPostgreSQL")
library(RPostgreSQL)
#a driver is needed
drv <- dbDriver("PostgreSQL")
#establish a connection
con <- dbConnect(drv,
                 user = "postgres",
                 dbname = "databaseName",
                 host = "myhost.com")
```

As a rule for all databases, the equivalent of SQL NULL in R is NA. The predicate `is.na` is equivalent to `IS NULL`; `!is.na` is the equivalent of `IS NOT NULL`. To get all rows without nulls in a data frame `df`, R uses `na.omit(df)`.

6.1.2 dbplyr

The `dbplyr` package can be seen as an extension of the popular `dplyr` package. `dbplyr` adds to `dplyr` the ability to deal with data in a database. This is helpful when the data is in the database to start with, or when the data does not fit into memory (more on this later). `dbplyr` connects to the database using DBI connections; for instance, it can connect to MySQL using the `RMySQL` driver and to Postgres using the `RPostgreSQL` driver. After the connection is established, `dbplyr` uses its own commands to move data between R and the database.

To start with, the package must be installed like any R package.

```
install.packages(c("dplyr", "dbplyr"))
library(dplyr)
library(dbplyr)
```

Then a connection to a database is established:

```
#create an in-memory SQLite database and copy over a dataset:
>con <- DBI::dbConnect(RSQLite::SQLite(), ":memory:")

#For databases not in memory, in a remote server:
con <- DBI::dbConnect(RMySQL::MySQL(), host = "hostname",
                user = "username", password = "myPassword")
```

dbplyr can be used with

- data in R: a data frame can be copied to a database table.
- data in files: data from files can be read into R and then copied to the database.
- data already in the database.

```
#copy from data frame to database table
copy_to(con, nycflights13::flights, "flights")

#from files:
df <- read_csv("filename.csv")

#Create a new SQLite database.
#the argument is a file that SQLite uses for its data
#Warning: any data in the argument is overwritten!
my_db_file <- "myfile.sqlite"
my_db <- src_sqlite(my_db_file, create = TRUE)

copy_to(my_db, df)
```

Data in the database is accessed using the `tbl()` command. The data can be retrieved in two ways:

- using SQL: dbplyr can be used to send SQL directly to the database. When a SELECT statement is used, the result is translated into an R data structure, as before.

  ```
  tbl(con, sql("SELECT mpg, wt FROM mtcars"))
  ```

- using dplyr commands: dbplyr will translate R code into SQL, send it to the database, get the answer back, and translate the answer into an R data structure (data frame). This option is detailed in the next example.

Example: Using dbplyr with R Commands

```
flights_db <- tbl(con, "flights")

flights_db %>% select(year:day, dep_delay, arr_delay)

flights_db %>% filter(dep_delay > 240)
```

```
flights_db %>%
  group_by(dest) %>%
  summarise(delay = mean(dep_time))

tailnum_delay_db <- flights_db %>%
  group_by(tailnum) %>%
  summarise(
    delay = mean(arr_delay),
    n = n()
  ) %>%
  arrange(desc(delay)) %>%
  filter(n > 100)
```

dbplyr translates the above code into SQL. The translation can be inspected by using show_query():

```
summary <- mtcars2 %>%
   group_by(cyl) %>%
   summarise(mpg = mean(mpg, na.rm = TRUE)) %>%
   arrange(desc(mpg))

summary %>% show_query()

<SQL>
SELECT 'cyl', avg('mpg') AS 'mpg'
FROM 'mtcars'
GROUP BY 'cyl'
ORDER BY 'mpg' DESC
```

Roughly speaking, the translation works as follows:

- head generates a LIMIT clause.
- filter results in conditions in a WHERE clause.
- select gives attributes in a SELECT clause (if no select is used, all attributes available are retrieved with '*').
- groupby gives attributes for a GROUP BY clause. This can used with summarise to compute aggregates. aggregates can be used for aggregates on their own (for instance, aggregates (n_distinct) generates COUNT (DISTINCT)) in SQL).
- arrange generates an ORDER BY clause.
- For dealing with multiple tables, dplyr has an "inner_join" function, as well as a "left_join" and righ_join functions that simulate a left (right) outer join (see Sect. 5.1). They take the attribute tables used for the join condition as arguments.[2]
- intersect, union, and setdiff generate INTERSECT, UNION, and EXCEPT in SQL.

[2]Note that only *equi-joins* are supported, but these are by far the most common case.

As shown above, to write the query in dplyr, you connect multiple commands using the %>% pipe to build a pipeline.

The system can also deal with functions:

```
mf <- memdb_frame(x = 1, y = 2)

mf %>%
  mutate(
    a = y * x,
    b = a ^ 2,
  ) %>%
  show_query()
#> <SQL>
#> SELECT 'x', 'y', 'a', POWER('a', 2.0) AS 'b'
#> FROM (SELECT 'x', 'y', 'y' * 'x' AS 'a'
#> FROM 'dbplyr_002')
```

It is highly instructive to build pipelines in dplyr and then show their SQL counterpart.

One of the advantages of using dbplyr is its ability to handle large dataset using a technique called *lazy evaluation*. Basically, when the user sends a query to the database, the system does not execute the query right away; hence, no results are generated (yet). When the user wants the result, she must ask for them, using collect():

```
mtcars2 %>% select(mpg, wt) %>% collect()

#Or, if a previous query was saved in a variable:
summary %>% collect()
```

Also, the system does not retrieve all results at once. This is why, when looking at some results, one may receive an answer with some data and a last line like

```
# ... with more rows
```

The system does not retrieve all results at once. This is also the reason that, when asking about the size of an answer, the system claims it does not know it:

```
nrow(mtcars2)

#> [1] NA
```

The user can ask for more results using collect(). While this may seem like a strange technique, it allows R to deal with results that exceed the size of the available memory.

6.1.3 *sqldf*

sqldf is an R package that also allows an R user to use SQL over R data or database data. sqldf can work with a data frame in R's memory; with data loaded from a file, and with data that resides in a database. After getting installed as usual in R,

```
install.packages("sqldf")
library(sqldf)
```

the package can be used with data frames simply by using the data frame name as a table name in the FROM clause of a query, so that the data frame can be manipulated using SQL:

```
data(mtcars)

sqldf("SELECT * FROM mtcars WHERE mpg > 20")
```

`sqldf` can also be used with a data file by using the `file` command; this creates a connection, which can then be used as a data source:

```
iris = file("iris.dat")
sqldf("SELECT count(*) FROM iris")
```

Another way to read from a file is to use the function `read.csv.sql`. This function, like `read.csv`, will take a file name to read as argument, but it will also take an additional argument called `sql`, where the user can put an SQL SELECT statement. Using this statement, it is possible to control exactly which data/how much data is read, something very useful when the file is very large.

We can perform insert, update, and delete statement through `sqldf`. However, it is important to understand that `sqldf` automatically copies the data from the data frame to the SQLite database, and all changes are made to the copy in the database, not to the data in R. To distinguish between the two, the copy in the database uses the prefix `main` in its name:

```
sqldf("INSERT INTO main.iris
            values(4.1, 3.2, 4.5, 1.3, 'setosa')")
data frame with 0 columns and 0 rows
sqldf("SELECT count(*) FROM main.iris")
  count(*)
1      151
```

The result returned by a `sqldf` query is a data frame, which can be saved and then used for further work. As a result, more complex analyses can be broken down into steps, just like using subqueries or views in SQL:

```
df <- sqldf("SELECT * FROM mtcars WHERE mpg > 20",
            row.names=TRUE)

sqldf("SELECT avg(mpg) avg, min(mpg) min, max(mpg) max
        FROM df WHERE make_model like 'Merc%'")
```

Example: Using sqldf

The following examples show how many data frame operations in R have counterparts in SQL and vice versa. Each pair of statements (the first one operating directly on the data frame, the second one using `sqldf`) produces the same result. Note: when comparing the results from R and SQL in this example, the user must take into account that the row names with the name of each car are not standard columns and will not be returned by sqldf unless the user asks for it, using `row.names=TRUE`:

```
sqldf("SELECT * FROM mtcars WHERE mpg > 20", row.names=TRUE)
```

It is also possible to make the row names into a regular column in the data frame by using

```
mtcars$make_model<-row.names(mtcars)
```

This will make comparing SQL and data frame manipulations easier.

```
#picking attributes
mycar = mtcars[,c("mpg","cyl","make")]
sqldf("SELECT  mpg, cyl, make FROM mtcars")

# head
head(mtcars)
sqldf("SELECT * FROM iris limit 6", row.names=TRUE)

head(mtcars, n=10)
sqldf("SELECT * FROM mtcars limit 10", row.names= TRUE)

# order
head(mtcars[order(mtcars$mpg, decreasing=TRUE),], 5)
sqldf("SELECT * FROM mtcars ORDER BY mpg desc limit 5",
       row.names = TRUE)

# subset
subset(mtcars, mpg > 25)
sqldf("SELECT * FROM mtcars WHERE mpg > 25", row.names=TRUE)

subset(mtcars, mpg > 25 & cyl < 5)
sqldf("SELECT * FROM mtcars WHERE mpg > 25 and cyl < 5",
      row.names=TRUE)

subset(mtcars, grepl("Mazda", mtcars$name))
sqldf('SELECT * FROM mtcars WHERE name LIKE "Mazda%"',
      row.names=TRUE)

# aggregate and GROUP BY
aggregate(mtcars$mpg, list(mtcars$cyl), mean)
r2= sqldf("SELECT cyl, avg(mpg) as avgmpg FROM mtcars
          GROUP BY cyl")
```

```
#subqueries, aggregated.
subset(mtcars, mpg > ave(mpg, cyl, FUN = mean))
#reusing previous result
sqldf("SELECT * FROM mtcars, r2 WHERE mtcars.cyl = r2.cyl
       and mpg > avgmpg")
#same query in single statement
sqldf("SELECT *
       FROM mtcars,
            (SELECT cyl, avg(mpg) as avgmpg FROM mtcars
             GROUP BY cyl) as t
       WHERE mtcars.cyl =t.cyl and mpg > avgmpg")

# table for pivoting
table(mtcars$mpg, mtcars$cyl)
sqldf("SELECT sum(cyl=4), sum(cyl=6), sum(cyl=8)
       FROM mtcars
       GROUP BY mpg")

# reshape
mycar = mtcars[,c("mpg","cyl","make")]
reshape(mycar, timevar ="cyl", idvar="make", v.names="mpg",
        direction="wide")
#this would take us back to the original dataset
reshape(wd, direction="long")

sqldf("SELECT make,
          sum(case when(cyl==4) then mpg else 0 end) as 'c4',
          sum(case when (cyl==6) then mpg else 0 end) as 'c6',
          sum(case when (cyl==8) then mpg else 0 end) as 'c8'
      FROM mycar GROUP BY make")
```

Other operations also have natural counterparts: assume df1 and df2 are two data frames with the same schema; then

```
# rbind
rbind(df1, df2)
sqldf("SELECT * FROM df1 union all SELECT * FROM df2")
```

produce equivalent result. Also, if df3 and df4 are two data frames (with arbitrary schemas),

```
# merge.
merge(A, B)
sqldf("SELECT * FROM A, B")
```

also produce equivalent results

6.1.4 Packages: Advanced Data Analysis

One of the reasons for the popularity of R is that it has a large number of commands that are specifically designed for data exploration, cleaning, pre-processing, and analysis tasks. Thus, actions that require a whole query in SQL (sometimes a complex one) can be done in one line in R. Also, R is adept at generating simple graphical displays for data, which can be very effective in EDA.

The command `summary`, with a dataset as argument, will make the system generate basic statistics about the input. It can also be applied to the result of any analysis.

Histograms are generated by the function `hist`. The partition of the beans can be specified by passing a vector of values as a second, optional argument: a vector with values $(a_0, \ldots, a_k)$ will create bins (a_i, a_{i+1}) for $i = 0, .., k - 1$. A boxplot can be generated by the function `boxplot`, which takes 1 or several attributes as parameters or even a whole dataset (categorical attributes are turned into numerical by enumerating them, which does not always make sense).

Plots are generated by the `plot` command: with two arguments, each an attribute, it will generate the scatterplot of both attributes. With a complex dataset as an argument, it will generate all scatterplots, one for each pair of attributes in the dataset. If an attribute is categorical, the function `as.numeric(attr)` will transform it into a numerical attribute for the plot. Jitter can be added to the scatterplot by calling `plot(jitter(attr))`.

A distance matrix for a dataset can be calculated with `dist(dataset)`. Correlation coefficients can be calculated as follows: assume `attr1` and `attr2` are two attributes. Then `cor.test(attr1,attr2,method="spearman")` will calculate the Spearman coefficient (this can also be used for "kendall" and "pearson").

Finally, complex analysis that was not doable in R without some serious programming can be accomplished using *packages*. A package is a file containing functions defined in R for some specific purpose. For instance, there is a library `outliers` that has some test of outliers. The available packages are too many to mention; the interested reader is referred to https://cran.r-project.org/web/packages and to https://rstudio.com/products/rpackages/ for more information.

6.2 SQL and Python

Python is an all-purpose programming language that has become extremely popular within the data analysis community, due to its built-in facilities to deal with datasets and to the large number of libraries developed to do sophisticated data analysis within a Python program. In this section we show how Python programs can be designed to interact with databases.

6.2.1 *Python and Databases: DB-API*

To read data from a relational database from Python, there is a direct connection using the DB-API, a collection of functions that can be used in any Python program to bring data from the database to the program, manipulate said data, and store it (or other results) back in the database.

Since Python is an object-oriented language (at least in its current version), the DB-API works by creating objects and using the methods attached to those objects. Two types of objects are used: *connection* objects and *cursor* objects. A DB-API access works as follows:

1. first, the program must establish a connection to the database. This is achieved through the `connect` method that takes in a URL, a user name, and a password and returns a connection object. The method establishes a connection with the database server at the given URL, as a user with the credentials provided.
2. the connection object can be used to create a cursor object, with the `cursor()` method. The cursor object is the one that will be used to send and execute SQL commands.
3. The cursor object has an `execute` method that takes in as parameter a string representing an SQL command. This command is sent to the database server and executed.
4. When the command is an SQL SELECT, a result is returned to the cursor object. This result can be retrieved with a variety of `fetch` commands:

   - `fetchall()` will take the whole result and deposit it in a Python variable of type array, with each element of the array representing a record and being itself an array. A typical access with this method will use `fetchall()` to deposit the result of a query in a Python list and then iterate over the list with a `for` loop, doing whatever is needed on each row (see example below). However, this can be quite slow if the result returned from the database is very large.
   - `fetchone()` will pick a row of the result at random. The typical approach is to call `fetchone()` inside a loop. This works for large datasets but makes access row-based. When using this function, the value `None` is returned when no more rows are left; a typical loop (assuming variable `cur` is the cursor object) is:

   ```
   while True:
       record = cur.fetchone()
       if record == None:
           break
       .... action on the retrieved row ...
   ```

   - An intermediate approach is to use `fetchmany(size=n)` where n tuples are returned at once (when there are fewer than n rows left, the method simply returns; however, many rows are left). This method is also used inside a loop but can be more efficient than `fetchone()`.

Calls to all these methods can be combined; whenever any `fetch` method is used, it consumes a certain number of tuples and subsequent calls consume the rest of the tuples. The cursor itself keeps track of which tuples have been consumed.

Example: MySQL Access with Python

```
import MySQLdb
#create a connection with database
conn = MySQLdb.connect('DatabaseName', user='username',
                       password='password')

#create a cursor object to handle interaction with database
cursor = conn.cursor()

#provide an SQL statement to be executed (note the quotes)
cursor.execute('''SELECT * FROM Table
                  WHERE attribute = 'value''')

#get the result of the previous query
result = cursor.fetchall()

#iterate over the result, one row at a time
for row in results:
      firstAttribute = row[0]
      secondAttribute = row[1]
      thirdAttribute = row[2]
#manipulated retrieved results as needed
      print "first attribute=%s,
             second attribute =%s,
             third attribute = %d" % \
         (firstAttribute, secondAttribute, thirdAttribute)

#close the cursor when finished
cursor.close()

#close the connection when finished
conn.close()
```

The cursor object has an attribute, `description`, that contains metadata—in particular, after a query, this is basically the schema of the returned answer; it is `None` for other commands. The description is a sequence of attribute descriptions; each description contains the name, type, and other information of the attribute.

```
cursor.execute("SELECT * FROM TABLE;")

# metadata from query
columns = cursor.description
```

```
# iterate over attributes
for column in columns
    print "attribute name =%s, attribute type =%d, %
    (column[0], column[1])
```

Another attribute, rowcount, tells us how many rows are returned by a query (or how many rows are affected by an INSERT or UPDATE).

Other accesses, besides queries, are also possible:

```
#create a database
cursor.execute('Create database Example')

#use that database as the current one
cursor.execute('use Example')

#create a table in database
cursor.execute('create table TName(
                          first-att char(30) primary key,
                          second-att int))')

#put some data into table
cursor.execute('insert TName values (%s %d)', ('foo', 30))

#make sure the insertion is in the database
conn.commit()
```

One nice characteristic of DB-API is that accesses to any relational system is pretty much the same, as can be seen comparing the previous example (which used a MySQL database) to the next one (which uses a Postgres database).

Example: Postgres Access with Python

```
import psycopg2 as dbapi2

#establish a connection
db = dbapi2.connect (database="dbname", user="username",
                     password="password")

# get a cursor from the connection
cur = db.cursor()

#use the cursor to send a query
cur.execute ("SELECT * FROM TableName");

# get the result from the query
rows = cur.fetchall()

#iterate over the result
for i, row in enumerate(rows):
    print "Row ", i, "value = ", row
```

```
#close the cursor once done
cur.close()

#close the connection once done
db.close()
```

Besides a result, an `execute` command returns a status variable telling the program if there has been an error. Hence, it is good programming style to wrap a call `execute` within a `try` block:

```
try:
   # Execute the SQL command
   cursor.execute(sql)
   # only if the sql command included changes,
   db.commit()
except:
   # only if the SQL command included changes,
   db.rollback()
```

With SELECT statements, no commit is needed (and if there is an error, no rollback is needed) as no changes are made to the database while querying. However, it is still a good idea to check the status variable before looping over the result since, if something went wrong, there is no result to loop over.

There is also a module to connect to a SQLite database.

Example: Using SQLite

```
import sqlite3
conn = sqlite3.connect('example.db')

c = conn.cursor()
# Create table
c.execute('''CREATE TABLE mytable (id int, name text)''')
# Insert data
c.execute("INSERT INTO mytable VALUES (1, 'Jones')")
c.execute("INSERT INTO mytable VALUES (2, 'Smith')")
c.execute("INSERT INTO mytable VALUES (3, 'Lewis')")
# Save  the changes
conn.commit()

for row in c.execute('SELECT * FROM mytable WHERE id > 1'):
    print(row)
```

It is also possible to create an array of rows, say `records` and call a single statement to do multiple insertions at once, using method `executemany`:

```
c.executemany('INSERT INTO mytable VALUES (?,?)', records)
```

Besides using DB-API, there are several other approaches, like `alchemy` and `django`. For more information, consult
https://docs.python-guide.org/scenarios/db/.

6.2.2 *Libraries and Further Analysis*

A common use of Python is to get some data from the database and carry out some tasks for which the database is ill-suited, in particular visualizations and complex analysis. This is achieved in Python through the use of *libraries*, collections of functions written by a programmer with a specific purpose in mind.

Some of the most common and useful libraries include:

- *NumPy* (Numerical Python), which provides operations on n-dimensional arrays and matrices, including many numerical routines on which other libraries and modules build;
- *SciPy* (Scientific Python), which builds upon NumPy and contains modules for linear algebra, optimization, integration, and statistics.
- *Pandas*, a package designed to define series (one-dimensional tables: each row is indexed and contains a single value) and data frames (two-dimensional tables: each row is indexed but contains several values) and provides tools for manipulation, wrangling, aggregating, and analyzing both of them.
- *SciKit-Learn*, built on top of SciPy and offering an array of Machine Learning capabilities.
- *NLTK* (Natural Language Toolkit), a library that provides tools for Natural Language analysis, including complex processing like tagging, text classification, named entity recognition, and parsing.
- *Gensim* is another natural language analysis library, but it focuses on sophisticated Information Retrieval methods like Latent Dirichlet Allocation (LDA) and distributed semantic representations (word2vec).
- *Matplotlib*, which provides multiple tools for data visualization.
- *Bokeh*, another data visualization library, specialized in interactive graphics.

There are many more libraries available.

Note that many of these assume that their data is in a certain format (tabular, most of the time) and that it has been cleaned and corrected. Thus, before using any of the functions on any of these libraries, it is a good idea to examine the data and make sure that it is clean and in the expected format, either using SQL or Python code to carry out data pre-processing.

Appendix A
Getting Started

A.1 Downloading and Installing Postgres and MySQL

To download and install Postgres, go to https://www.postgresql.org/download/. In this page, there are a variety of options for different operating systems. Postgres is available for Windows, Mac OS, and Unix/Linux. The windows version has an installer that downloads and installs the database server as well as pgAdmin, a well-known and commonly used GUI (Graphical User Interface). Likewise, Mac OS also has an installer that downloads and installs a bundle with the server and pgAdmin on it. On top of that, Postgres is the default database for Mac OS since version 10.7 and so a version may already be installed on your Mac. However, this version may not be the latest available. There are other ways to download and install Postgres, all of them described at https://www.postgresql.org/download/macosx/.

To download and install MySQL, go to http://dev.mysql.com/download. In this page, you will see a list of options. You want to use the link to the MySQL Community Server, which is the free database server we have been using in this book.[1] On the page that the Community Server links to, you are given a set of choices depending on the operating systems your computer runs. MySQL has versions for Windows, Mac OS, and Unix/Linux. If you choose the Windows version, you will be recommended to use the MySQL Installer for Windows. If you choose this option (and you should, unless you know what you are doing) you will be taken to another page where there will be two choices: a 'full' installer that downloads about 400+ Mbs for a bundle that contains the server and a bunch of other

[1]MySQL AB, the company behind the software, was bought by Oracle, a large and famous database software company, in 2008. However, the Community Server version remains open-source and free of charge. Another open source and free database system, MariaDB, has been developed from MySQL and independently of Oracle. You should be able to download and install MariaDB from https://mariadb.com. The MariaDB Community Server is a highly compatible replacement for MySQL and has new features added periodically.

A. Badia, *SQL for Data Science*, Data-Centric Systems and Applications,
https://doi.org/10.1007/978-3-030-57592-2

packages, and the 'Web' installer that only downloads about 25 Mbs. This package allows the user to choose which pieces of software to install. Only the server is needed if one is going to use the CLI (Command Line Interface) to interact with the database. However, it is recommended (especially for first-timers) that a GUI (Graphical User Interface) is also installed (see next section). The most popular GUI for MySQL is MySQL Workbench.

On either systems, as installation proceeds, some dialog boxes will come up. Most of the time, the default options are fine. There is one thing that you should pay attention to: when the database is installed, a first user, called the *superuser*, is created (username in Postgres: `postgres`; in MySQL: `root`). The system will ask for a password for this superuser. It is important that you do not forget this password, since in order to start using the database you will need to connect with the superuser's credentials until more users are created (see Sect. A.3).

A.2 Getting the Server Started

The software you just downloaded and installed contains two vital pieces, called the *server* and the *client*. The server is the part that will take care of storing the data, modifying it, and accessing it. It executes all SQL commands. The client is the part that interacts with the user: it issues a prompt on the screen (when used in command line mode, see below), takes a command from the user, sends it to the server, gets an answer back from the server and displays it for the user, and issues the prompt again. Clearly, server and client must talk to each other: this is called a *connection*. Using a connection, the client sends the server the user commands; the server executes them and gives the client an answer (and also error messages, if anything went wrong). A server can talk to several clients at once; this makes it possible for whole teams of people to access and manipulate the same data, in the same database, at the same time.

Many times, server and client both run in the same computer. However, they can run on different computers; sometimes, a server is installed in a large, powerful computer so it can handle large amounts of data and many clients, while the client is installed in a personal computer. In this case, communication takes place using a network. That is why sometimes when trying to establish a connection, it is necessary to provide the client with a *host* name: this is the name of the computer where the server is (when the server is in the same computer as the client, the name 'localhost' is sometimes used).

Because the server must be ready to accept requests from a client at any time, the server is a program that runs continuously; when no client is contacting it, the server is simply waiting for some client to start a connection. Thus, after installing database software, the first thing to do is to start running the server. In some systems, this is done automatically: the system realizes it is a database server that is being installed, so as soon as it is ready, the system gets the server running and ready to receive connections. By contrast, a client need not be active until a user actually needs to work with the database. The user activates the client and then uses it to establish a connection with the server.

Before accessing a database, a user must be registered with the database. As indicated above, when a server is installed, a first user is created by default. After starting the server (if not started by default), it is necessary to establish a connection with the database as the default user and create other users so that they can access the database. This is done with a specific SQL command—user management is described in the next section.

There are two types of clients: CLI (command line based) and GUI (Graphical User Interface) based. A command line client is activated by typing the command name in a terminal (the name is `mysql` for MySQL and `psql` for Postgres). With the command, it is typically necessary to give several arguments, including a username and a password as well as a hostname (see above). A GUI client is a separate program that offers a window-based environment to manage connections to the server. There are several free GUI based clients for both MySQL and Postgres; the most popular for MySQL seems to be MySQL Workbench (pictured in Fig. A.1), while the most popular for Postgres is probably pgAdmin (pictured in Fig. A.2). Both can be downloaded and installed in a personal computer from the same sites from which the database server was obtained. When starting a GUI based client, a window pops us and offers the user a set of menus. A connection is started by choosing “Add a server” (an option under the File menu) and then clicking on the Add a Connection icon (in pgAdmin) or by choosing “Connect to a Database” from the Database menu (in Workbench). In either case, a hostname, a database name (the name given in the CREATE DATABASE or CREATE SCHEMA command), a username, and a password must be provided (it is possible to set defaults and/or

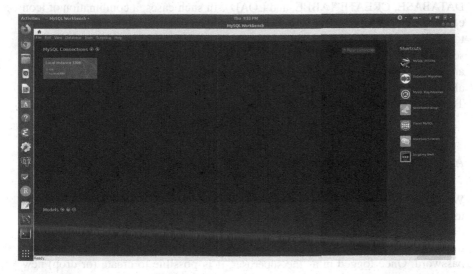

Fig. A.1 A screenshot of MySQL Workbench

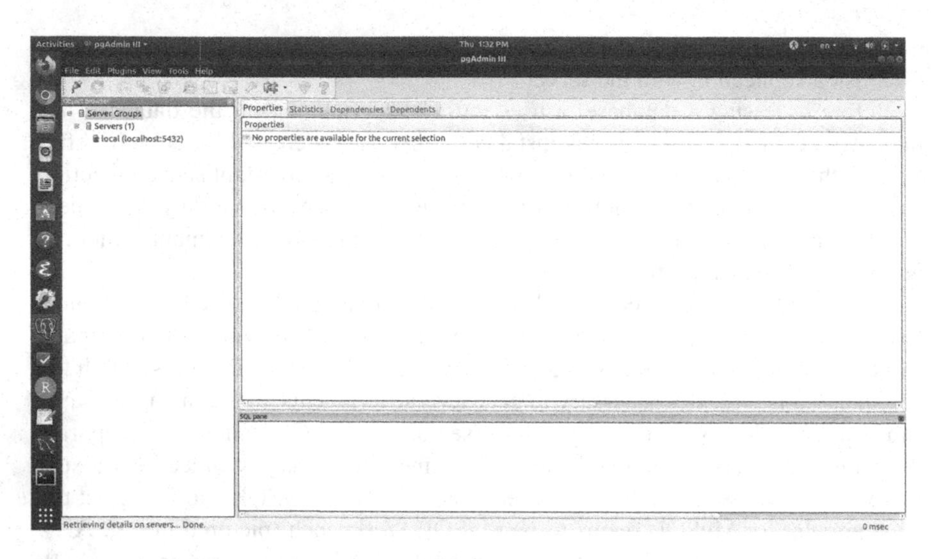

Fig. A.2 A screenshot of Postgres pgAdmin

have the system remember some of those). In most systems, after a connection to a server is established for the first time, it can be saved and an icon created for it. Clicking on it will start the connection without the need to retype anything but the password.

Commands are always sent to the database in SQL. In the command line, the user types SQL; in the GUI, there is always an SQL pane where the user can type SQL, but there are also shortcuts for some common operations (like CREATE DATABASE, CREATE TABLE, and LOAD). In such cases, a combination of icon clicking and filling in some forms will get the job done. However, it is important to emphasize that even when these shortcuts are used, the communication with the server is carried out in SQL. What the GUI client does is to take the user clicks and choices and compose SQL commands automatically (in Workbench, the user is always given the option to check the generated SQL before sending it to the server). Therefore, what we have shown in this book can be used with either type of client.

A.3 User Management

When connecting to the database server, the connection is always as a database user. A user is identified by a *username* and a *password*. As stated earlier, when the database is first installed, a 'superuser' is created. This makes it possible to access the database for the first time; the user simply uses the superuser's username and password. Once logged in as the superuser, it is possible to create (or drop) new users for the database. After a user is created, anyone who knows the username and password of that user can connect to the database as that user.

Unfortunately, user specification is not part of the SQL standard, so each system does it a bit differently. However, most systems have coalesced around a set of very similar operations, so once one is familiar with a given system, it is usually not that different to learn a new one.

Postgres uses the concept of a *role*, which is a user or group of users with similar permissions. In Postgres, the SQL command to create a user is

```
CREATE ROLE rolename WITH PASSWORD password;
```

or, equivalently,

```
CREATE USER username WITH LOGIN PASSWORD password;
```

There are also DROP ROLE and ALTER ROLE commands.

Associated with each user is a series of *permissions* (also called *privileges*) that tell the system what the user is authorized to do. To give a user permissions on a database, in Postgres one must start by allowing the user to connect to the database:

```
GRANT CONNECT ON DATABASE database TO username;
```

Then, access to schemas in the database is granted with

```
GRANT USAGE ON SCHEMA schema-name TO username;
```

After that, permissions can be granted:

```
GRANT permission-list ON TABLE table-list TO username;
```

The permission-list is a list of the actions that a user is authorized to carry out in the database. These include SELECT, INSERT, DELETE, UPDATE, CREATE, DROP, ALTER (the last three refer to tables). To give someone blank permission to do anything, the keyword ALL is used.

The table-list is a list of tables in the schema; the expression
ALL TABLES IN SCHEMA schema-name
can be used to give permission in all tables of the given schema.

A common procedure in Postgres is to create a role, assign this role permissions using the GRANT commands shown, and then create individual users and assigning roles to users. Each user inherits all permissions from the role. The procedure looks like this:

```
CREATE ROLE rolename;

GRANT CONNECT ON DATABASE database_name TO rolename;

GRANT USAGE ON SCHEMA schema_name TO rolename;

#Or, to allow permission to create tables:
GRANT USAGE, CREATE ON SCHEMA schema_name TO rolename;

GRANT SELECT, INSERT, UPDATE, DELETE ON ALL TABLES
```

```
IN SCHEMA schema_name TO rolename;

CREATE USER username WITH PASSWORD 'user_password';
GRANT rolename TO username;
```

To view all users in a server, in the command line simply type du. This will produce a table with schema (Role-name, attributes), that will display the permissions per role (command du+ provides additional detail, while command du username provides information about a given user). This is actually a shortcut for an SQL command that looks into a table called pg_catalog.pg_user, which stores information about roles/users.

To change existing permissions on a user, use

ALTER USER permission-list IN database

where the permission-list can be ALL or a list of allowed actions (as above). To change the name of a user, the shortcut

ALTER USER old-username RENAME TO new-username

is provided. To take away some privileges, the REVOKE command is used:

REVOKE permission-list ON database FROM username;

To delete a user, simply use

DROP USER username;

If roles were used instead, one can drop the role instead:

DROP ROLE rolename;

In MySQL, the SQL command to create a user is

```
CREATE USER username IDENTIFIED BY password;
```

The username is sometimes written as username@host, where host indicates the computer from which the user is expected to connect to the database. For a database running in the same computer as the client, localhost is used.

Of course, there is also a DROP USER username to delete a user and ALTER USER username to make changes to a user. The most common changes are to change the username and/or the password or to adjust the permissions of a user.

The command to assign permissions to a user is

GRANT permission-list ON database TO username

The permission-list works as in Postgres. The database indicates on which databases the permissions are granted (since a server may run several databases); to give permissions in all the databases in a server, the '*' symbol can be used. If one wants to give permissions on only some of the tables of the database, dot notation (database.tablename) is used to indicate so.

Finally, username identifies the user to which the permissions are given.

The most important aspects to know about permissions are:

- permissions can be roughly divided into *read* and *write* permissions. Read permissions allow a user to 'see' things and are needed to run queries (i.e. the SELECT permission). It is common to create users with read permission over all or a part of the tables in a database, which will allow those users to run queries over such tables. Write permissions are needed to 'do' things, including changing existing things (i.e. the INSERT, DELETE, UPDATE, CREATE, DROP, ALTER). Thus, write permissions are needed to insert, delete, or update data in tables and also to create, drop, or alter tables. Obviously, users who need to upload data into the database (or move it out) need to write permissions.

- a user may be given permission to create other users (a specific CREATE USER permission). In this case, a separate GRANT OPTION permission allows the user to pass some or all of its privileges to the newly created user. If user A is given the permission to create another user B, it is typically the case that A cannot give B permissions that are more powerful than its own. This prevents A from creating a user B with all-encompassing permissions and then (knowing B's username and password) log into the system as B and do things that it (A) was not allowed to do in the first place.

The original user is called a 'superuser' because it has permissions to do anything—otherwise, there would be something that no user could ever do, since the superuser is the origin of all other accounts. Some system administrators reserve the superuser account for themselves and make all other database users to be 'ordinary' users, with limited permissions.

To view all users in a server, type

```
SELECT User, Host FROM mysql.user;
```

To see what kind of permissions a user has, type

```
SHOW GRANTS FOR username;
```

To remove privileges, use the same REVOKE command as Postgres.

A shortcut to change a user's username is to use RENAME:

```
RENAME USER old-username TO new-username;
```

and to change the password

```
SET PASSWORD FOR username = PASSWORD(password);
```

Appendix B
Big Data

B.1 What Is Big Data?

In recent years, the amount of data available for analysis has increased tremendously in many fields. This is due to the confluence of several factors: more processes are mediated by computer (think online shopping, automated factory control, etc.) and it is very easy to have the computer keep a record of every single operation it carries out; storage is getting cheaper, and software tools for handling these large datasets have been developed.

As a consequence, many people talk about *Big Data* to refer to this increase in the volume of data. However, this is a very imprecise term that means different things to different people in different contexts. When does data become *big*? Is it just a question of size?

Sometimes people refer to the five (or seven or even ten) Vs of Big Data; the basic three that everyone seems to agree on are:

- Volume: this refers to sheer size. If data is tabular, this usually means the total number of rows in the tables in your database. Nowadays, one can find tables with tens of millions of rows—and more. This means that most analyses, even the simple types described in Chaps. 3 and 4 are challenging to implement in a fast (interactive) manner. This has led to considerable research, and several solutions to this challenge (which we outline below) are available.
- Variety: this refers to the diversity or heterogeneity of the data. When data comes from several sources and it is in several forms (some structured, some semistructured, some unstructured) it usually must be integrated, that is, put together under a common schema, before analysis. Such integration is extremely difficult; in spite of many years of research and many papers in the subject, integrating data is still more art than science, and it usually requires a large investment of time and resources.

A. Badia, *SQL for Data Science*, Data-Centric Systems and Applications,
https://doi.org/10.1007/978-3-030-57592-2

- Velocity: this refers to the rate at which data keeps on coming; you will sometimes hear people refer to *streaming* data or similar terms. If data keeps on coming at high speed, it not only increases rapidly in volume; it also puts the results of analysis at the risk of becoming obsolete quite rapidly. This creates its own challenges if we want to keep our information up-to-date.

However, most of the time when people use the term Big Data they refer to the first characteristic only, Volume. The question, then, is when is Big Data big enough to deserve this name? To explain where the border lies, we have to explain a little bit about computer architecture.

There are *two* distinct memories in any computer. The first one is sometimes called RAM, while the second one is usually referred to as the *disk* (or the drive or the hard drive).[1] The computer moves data and programs between these two memories because each one has a different function. RAM is used for data that the computer needs to operate on; it is very fast, but is also *volatile*—meaning that all the contents of RAM are erased if the computer is turned off or it crashes. The disk, however, is what is called a stable memory: it keeps the data whether the computer is on or off. It is used for long-term storage; that is where files reside. Unfortunately, the computer cannot work with the data in the disk; it needs to move data to RAM before it can make changes to it or even read it for analysis. Because of the current state of technology and manufacturing, disk is much cheaper than RAM, and so most computers have much larger disks than they have RAM memory—both RAM and disks have become cheaper with time, so computers come with larger and larger amounts of both, but the ratio of the size of RAM to the size of disk has not changed that much in a decade. As of 2019, a typical computer will have between 30 and 100 times more disk than RAM.

In summary, RAM is fast, unstable, and (relatively) small, while disk is slow, stable, and big. As a consequence, any activity that requires accessing large amounts of data from disk becomes too tedious and cumbersome; in some cases, it is not even doable. To cope with large data, database systems store data on disk but try to become efficient at accessing it, either by using quick access methods (indices) or by optimizing the execution of queries or a combination of both. They still need to move the data from disk to RAM, so access to large datasets may take a while. Sometimes the data to be analyzed may not fit into RAM; in those cases, database systems are designed so that they still can carry out the required task—at some extra time cost. This is what allows databases to handle very large amounts of data.

In contrast, R works with data in RAM. It reads data from a file in the disk, but it needs to be able to move all the data to RAM to work with it. As a result, R is limited on the size of datasets it can handle.[2] Python, as any program language, works with files and allows a programmer to decide exactly how to handle such files. Hence, Python can also deal with very large datasets, but it is up to the programmer to find

[1] In many modern laptops, hard drives are being substituted by other types of devices, so sometimes one may hear talk of SSDs instead.

[2] New packages are being developed to work around this limitation (see Chap. 6 for more on R).

ways to handle large files efficiently. Database systems, however, take care of this in a manner that it is transparent to the user.

Thus, the line for in-depth analysis is crossed as soon as data exceeds the size of available RAM. However, lightweight or selective analysis can still be done with data that fits on a medium-sized disk, although there will be an initial hit when moving the data.[3] One should be aware of the size of datasets that they need to handle and compare them to their computer's capability.[4] Fortunately, most people work with little (or medium-sized) data, not Big Data. The exceptions are people working at the so-called Internet companies (Google, Facebook, Twitter, Uber, etc.) or at large retailers (Walmart, Target) or at medium or large companies that happen to handle data as part of their regular business. The rest of us rarely will see the amounts of data that qualify as big.

Once this is said, people's jobs and requirements change, so we summarize here some technologies that help deal with large data. It is worth noting that this is an instance where relational databases shine. Because their data model and language are independent of the underlying hardware, it does not matter whether a database is running on a laptop, a computer in a local lab, a remote server, a computer cluster, or the cloud. The only thing a user needs to care about is learning how the relational model stores data and how to use SQL—something that hopefully this book has helped with. Once these skills are acquired, they can be deployed in a wide range of scenarios, as we will see.

B.2 Data Warehouses

A *data warehouse* is a special type of database that is designed to take in large amounts of data as far as such data fits into a certain model, sometimes called the *multidimensional* data model. Data warehouses have several techniques to support fast analysis of very large datasets. However, data warehouses are geared toward business analytics and are not necessarily a good match for general data analytics—in particular, Machine Learning mathematics-based computations. Since they support SQL, though, everything that was explained in this book can be done in a data warehouse and at a large scale.

The multidimensional model looks at data as a collection of basic *facts*; typically, there is only one or very few types of facts in a data warehouse. Each fact describes the basic activity that the warehouse was built to record. For instance, a data warehouse for an e-commerce website usually registers each single visit to the site:

---

[3] A modern disk can read data at anything between 80 and 160 Mbs per second, so a 1 GB file will take about 13 and 7 s to read, while a 100 GB file will take between 22 and 12 min.

[4] Another practical roadblock is *downloading* data. If the data must be obtained from the Web or a remote site, and downloading speed is 50 Mbs per second, 1 GBs of data will take approximately 21 s to download (assuming no network issues); 100 GBs will take about 35 min (and you can be almost assured of some network issues).

the page visited, the day and time, the IP address of the visitor, the OS and browser of the visiting computer (if available), and what was clicked on (if anything). Another typical example is a telecommunication warehouse that registers phone calls: the basic fact here is the call, including which phone number called, which phone number was called, the day and time, and the length of the conversation. Each fact is described by a series of characteristics, which are divided into *measures* and *dimensions*. Measures are typically simple numerical values (like the length of the phone call), while dimensions are complex attributes for which more information is available (like the phone numbers involved in a conversation, for which associated information may include the person having such number, her address, etc.). The data warehouse is organized as a relational database with one or more *fact tables*, which hold the basic information about the facts, and *dimensions tables*, one for each dimension, holding the information available for that dimension. For instance, in the telecommunications database we would have a `CALLS` fact table, with a schema like `(day, time, caller, callee, call_length)`. We would have at least one dimension table, `ACCOUNTS`, so that each `caller` and `callee` are associated with an entry on this table, which would include attributes like `(phone-number, name, address)` and possibly many others. Note that `(phone-number)` would be the primary key of the `ACCOUNTS` table, and that both `caller` and `callee` are foreign keys to it. This is a typical organization for a data warehouse, with the fact table holding (at least) one foreign key for each dimension. This is many times visualized with the fact table in the middle of a circle-like arrangement of dimension tables; because of this visual, this schema is typically called a *star schema*. As another example, the e-commerce site may have a fact table `VISITS`, registering each visit to a page, and may have dimensions `PAGE` (where the information contained on a web page is described) and `CUSTOMER` (if the visitor is a registered customer and has logged in to the site or can be identified based on IP address). In some data warehouses, the dimensions may contain complex information and therefore keeping all information about a dimension in a single table may lead to redundancy due to some of the reasons discussed in Sect. 2.2. In those cases, dimension tables are *normalized* and broken down into several auxiliary tables to avoid redundancy. When this is done, we talk about a *snowflake schema* instead of a star schema.

Note that in both the example of the e-commerce site and the telecommunications company, there is temporal information involved. Time is almost always one of the dimensions of the data warehouse, as the warehouse tends to accumulate collections of facts for a relatively long period of time (usually, a few years) in order to facilitate temporal analysis. Sometimes, the temporal dimension is explicitly stored in a table, especially when time must be broken down into periods that are not standard: days, weeks, and months can be considered standards, while others (like fiscal quarters) may not be. Sometimes, the temporal dimension is handling implicitly (in SQL, though data manipulations with date functions, without having an explicit table for it).

The fact table tends to be very large, as it registers a large collection of basic facts. Think of a retailer like Amazon, which does log (collect) all visits to all the pages in its site, when such visits may number in the millions each day; or a large phone company, which also handles millions of calls each day. It is not unknown to have fact tables with hundreds of millions of rows on it. By contrast, dimension tables tend to be considerably shorter, from hundreds to a few thousand rows.

Data warehouses typically get their data from other databases, usually called *transaction* databases because they handle transactions, that is, changes to the database in the form of INSERT statements that keep data in such databases current. The typical example is a brick-and-mortar retailer chain, which may have a database running on each store, registering the sales that are taking place in that store. Every so often, all data in the store database is transferred to data warehouse which collects all data for the whole chain—that is, all stores are connected to this data warehouse and transfer their data to the warehouse. This "gathering" of data is one of the reasons that the fact table in a warehouse grows so large (another one is that data, as stated before, is kept for quite some time).

The transfer of data between the transaction databases and the data warehouse is usually carried out at pre-determined intervals (typically, anything from 6 to 24 h is used). During this transfer, data from all sources is typically cleaned, standardized, and integrated, so that it all fits nicely in the schema of the warehouse. This is typically called the *ETL (Extract-Transform-Load)* process. When data comes from heterogeneous, highly variable sources, this can be quite a complex process which requires large amounts of code. In our example of a retail chain, chances are that all stores use the same schema for their databases—since it was likely designed by a group within the company for all stores to use. In such cases, ETL is much simpler.

Data warehouses are typically used by business analyst which take advantage of the fact that the warehouse contains data from the whole company to ask *global* questions (i.e. how are the stores in the Northwest area doing with respect to other areas? Or, which were the top 10 sellers throughout the whole company in the last week?). They also use the fact that the warehouse stores data for a long time to examine patterns of evolution and change (i.e. how are sales of perishable goods done over the last year? What was the month with the highest (and/or lowest) monthly sales of such items?). Clearly, such questions could not be asked of the database in any particular store.

The typical queries asked of a data warehouse are sometimes called *Decision Support* queries, and warehouses are sometimes referred to as *Decision Support Systems (DSS)*. The reason is that the information obtained through data warehouse queries are used by middle- and upper-level managers to make decisions about the company's future actions, i.e. purchases or discount campaigns. These queries are sometimes described using a specialized vocabulary (*slice and dice*, *roll-up*), but they are simply SQL queries that use certain patterns over the multidimensional data in the warehouse. To explain them, it is useful to visualize multidimensional data as points in a multidimensional space (just like we did in Sect. 4.3.1). As a simple example, assume a SALES fact table with dimensions PRODUCT, CUSTOMER, and STORE (i.e. what was sold, to whom, and on what store). We can see each record in SALES as representing a point in three-dimensional space: given a product p, a customer c, and a store s, the point at (p, c, s) is associated with certain measures (say, the number of products bought n and the total amount spent a).

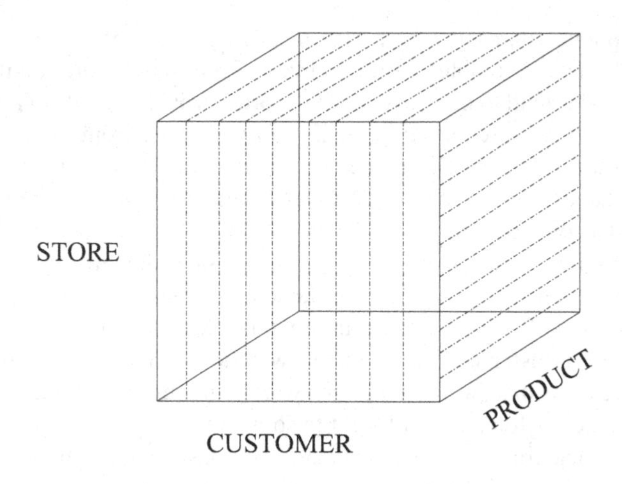

Fig. B.1 Three-dimensional data points as a cube

This point and its associated measures are represented the record `(p, c, s, n, a)` in fact table `SALES`. In a relational database, this would be implemented giving each dimension a primary key (say, `storeid` for `STORE`, `custid` for `CUSTOMER`, and `productid` for `PRODUCT`) and using them as foreign keys in the `SALES` table. We can also visualize the space as a cube since we conveniently made it three-dimensional (see Fig. B.1).

We can now describe typical data warehouse queries as follows:

- *Slicing*: to cut a slice of the cube, we take one of the dimensions and we fix it by using an equality condition. For instance, using `storeid = 112` cuts a slice of the cube along the `STORE` dimension. Slicing allows us to reduce the number of dimensions of what we are observing; it can be done using more than one dimension. Sometimes slicing refers to fixing an attribute of a dimension; for instance, assume that table `STORE` has attribute `city`; we use the condition `STORE.city = 'Lexington'` and join tables `STORE` and `SALES` in order to retrieve measures and other information related to sales in stores in that city.
- *Dicing*: to dice up the cube, that is, to get a smaller cube of the larger one. This is usually done by selecting on one or more dimensions with a range condition, that is, one that obtains a range of values instead of a single one, or with a condition on a categorical attribute. As with slicing, this can be done on the dimension itself or on an attribute of the dimension. For instance, using `PRODUCT.price BETWEEN 10 and 100` would take a 'chunk' of the `PRODUCT` dimension; using `STORE.state = 'NY'` would do the same (assuming there are several stores in the same state).
- *Roll-up*: facts in a fact table are usually not interesting on their own; what the analyst is really after is any trends and patterns in the collection of facts. Hence, when facts (dimensions or measures) are analyzed, they are usually *grouped and summarized* (using GROUP BY and aggregates). This also results in a small (hence easy to interpret) answer—it is difficult to make sense of an answer with

hundreds of thousands of records. The typical approach aggregates facts by some characteristics of one or more dimensions; for instance, we could look at sales by each state. In many instances, we can look at categories at several levels, as the attributes are naturally organized in a hierarchy: for instance, geographical and temporal hierarchies are very common. Geographically, we can look at all sales in a city, a county (grouping several cities together), a state (grouping several counties together), a region (grouping several states together), or a country (grouping several regions together). Temporally, we could look at all sales on a day, a week, a month, a quarter, or a year. In such cases, to *roll-up* is to summarize facts by one of the low levels in the hierarchy and then move up the hierarchy, each time getting a smaller, more coarse answer (but ones where patterns may appear more clearly). For instance, we may start to look at sales per city, then roll up by county, and then by state. In SQL, this can be accomplished with `ROLLUP` and `CUBE`, as described in Sect. 5.3.

- *Drill down*: this is the opposite of roll-up; it consists of going down in a hierarchy to get more fine grained answers. Sometimes we may notice something in an answer and we may need more detail; for instance, when looking at sales per quarter, we notice that one of them has substantially higher sales than the others. We then drill down to the months making up this quarter, and we find one of them that stands out. We again drill down to the weeks of that month, and we find a particular week that explains most of the surge in sales.

In most data warehouse queries, the fact table is joined with one or more dimensions, conditions are placed on such dimensions, and then a grouping and several aggregates are applied. Moving across dimensions and conditions allows us to examine those facts from several angles.

The important lesson to remember is that, in spite of all the new vocabulary, ultimately the data warehouse is basically a specialized relational database and the analysis of its data is done using SQL. However, there are also many software products that connect to a data warehouse, extract some of the data from it, bring it to memory, and analyze it there in order to achieve fast response times. Many times, such products have their own query languages, usually with ad hoc constructs that allow simulating spreadsheet-like operations on the data. Unfortunately, since there are many such languages, and each one has its own idiosyncrasies and peculiarities, we will not be describing them here. Most of them do not do anything that cannot be done in SQL.

We close this very brief description by pointing out that, in some cases, data warehouses may have complex schemas, with more than one fact table and many dimension tables for each. Being a central repository of information for the enterprise, data warehouses can grow in scope, complexity, and size. In some cases, a group of users that are only interested in part of what is available in the warehouse create a *data mart*. This is a database that contains only part of the data in the warehouse, either by restricting the schema (usually, picking one of the fact tables if there are several, and a subset of dimension tables) or by restricting the data (usually, picking some of the data in the fact table; for instance, the facts for the last

6 months). A data mart simplifies analysis and allows for faster response time since it does not have the size of the data warehouses. Usually, one or more data marts may be created after a data warehouse is up and running to service the needs of users with interest in only some parts of the warehouse. Again, a data mart is essentially a relational database and is managed using SQL (although, as in the case of the data warehouse, specialized tools may be used with the data mart to facilitate analysis).

B.3 Cluster Databases

A *computer cluster* is a collection of computers that are connected to each other, usually by a high-speed network. Many times, the name cluster is reserved for computers that are physically close (in the same room or perhaps the same building), so that the network used is a fast local area network, which is very fast and under the control of the cluster. Each computer in the cluster is called a *node* and is a complete computing unit (i.e. with its own CPU, memory, and disk, running its own operating system independently of others). However, all computers in the cluster are run in coordination by a special software, usually with all computers in the cluster running the same task. For many processes, the cluster appears as one large computer.

Clusters are used to deal with large amounts of data. Usually, the data is distributed across all the nodes in the cluster. In the case of a relational database, the data is usually distributed by *horizontal partitioning*: the tables are broken into chunks, disjoint collections of records. All chunks for a table share the same schema with the original table. Computer in the cluster chunk is given to each one.

Database systems developed to run in clusters take SQL, just like a traditional database, but then internally break down the tasks needed to execute the query into pieces that are sent to each node. A node runs the piece received on its own chunk of data, perhaps exchanging data with other nodes if necessary at some point. Because all nodes work in parallel, and each one of them handles only a chunk of the whole dataset, clusters are able to handle extremely large datasets.

The best well-known example of open-source software that controls a cluster is Apache Hadoop.[5] Hadoop provides a set of facilities to control the cluster. For instance, Hadoop has tools to deal with *fault tolerance* and with *load balancing*. Fault tolerance refers to the problems caused by one of the nodes in the cluster going down or crashing. This could prevent the whole cluster from working if no provisos are made for this situation. Load balancing refers to the fact that, for a cluster to be as effective as possible, ideally all nodes should have about the same amount of data to deal with. An uneven distribution of the data could cause some nodes to struggle with a large dataset, while others are idle because they have little data to work on. Distributing the data in a balanced manner clearly helps performance, but is not a trivial task.

---

[5] https://hadoop.apache.org/.

Cluster databases support the analysis of large datasets by providing algorithms that execute most SQL queries using all the machines in the cluster. The work is automatically divided up so that all machines collaborate in obtaining an answer. This can be done automatically in SQL, thanks to its declarativeness and relative simplicity. The great advantage of this is that a user can use SQL for analysis and not worry about how the system will actually carry out the queries. Thus, most of what we have seen in Chaps. 3 and 4 can be used in these systems. The open-source data warehouse Apache Hive[6] has been developed on top of Hadoop. Hive supports most of the SQL standard; all the approaches explained in this book can be used with Hive.

Most commercial database systems (and a few open-source ones) have 'cluster' versions. Installing and maintaining these versions can get quite complicated due to the number of parameters that must be set. In many systems, it is necessary that the user decides which tables must be partitioned, how they should be partitioned, how many replicas (copies) should be kept, and so on. It is a goal of current research to make these tasks simpler by using algorithms (including machine learning algorithms) to simplify them. Still, given their added complexity, these services should only be used when the volume of data justifies it. As a rule of thumb, the volume of data should be considerably larger than what one can fit in the hard drive of a powerful computer. If a cluster is needed, it is probably a good idea for a data scientist to team up with a *data engineer*, someone who is knowledgeable about cluster deployment and maintenance.

As a brief example, the CREATE TABLE command in Apache Hive looks and behaves pretty much like the standard SQL CREATE TABLE defined in Sect. 2.1. However, it has some additional, optional parameters:

```
PARTITIONED BY (col_name data_type
CLUSTERED BY (col_name, col_name, ...)
SORTED BY (col_name [ASC|DESC], ...) INTO num_buckets BUCKETS
SKEWED BY (col_name, col_name, ...)
    ON ((col_value, col_value, ...),...)
```

All these refer to how the data is laid out in the cluster. The `PARTITIONED BY` indicates a way to partition and distribute the data in the table among nodes in the cluster (essentially, all records with the same values for the partition columns are sent to the same node). Within each node, the data can be *bucketed* using the `CLUSTERED BY` clause; again, all records with the same values for the attributes named in this clause are stored together. Within a bucket, records may be kept in sorted order according to the attributes mentioned in `SORTED BY`. To improve load balancing, Hive allows the user to specify attributes that are heavily skewed (i.e. some values appear very often, while some others appear sparsely). The system will use the information on `SKEWED BY` (which provides not only skewed attributes, but also a list of frequent values for each) to even out data distribution by splitting the collection of records associated with frequent values.

[6]http://hive.apache.org/.

The purpose of all these optional, extra parameters is to facilitate very fast retrieval of data. It is easy to see that all these extra parameters are only useful in the context of a cluster; that is why they are absent from the standard definition. It is also clear that, in order to make good use of them, the user must have detailed information about how the data is distributed—not all attributes are good candidates for `PARTITION BY`, `CLUSTERED BY`, or `SORTED BY`. That is why it is a good idea to work with a data engineer for these decisions. The good news is that all these are irrelevant for data analysis; in a `SELECT` statement, only the table name is needed in the `FROM` clause: queries can be written just as they would over a regular database—just as it has been described in this book.

B.4 The Cloud

As explained in the previous section, cluster computing is very powerful but it also requires quite a bit of expertise in computers to implement properly. This has discouraged users that are not computer experts from working with large datasets. Lately, an idea has come to the forefront for attacking this problem: the idea of *computing-as-a-service*. This idea is based on the following premise: someone (a "cloud provider") sets up one or several computer clusters and installs software on them (in particular, database systems). This set of clusters is referred to as *a cloud*. The cloud provider sells access to these capabilities to the public. A customer can buy access to a certain amount of resources (disk, memory, CPU). Once a customer creates an account, she can access the resources needed, usually through a Web-based interface. In particular, in the case of databases, once connected into the system the user can create a schema, create tables on it, upload data to this schema, and run queries on it. The cloud provider makes sure that everything runs smoothly. If the customer needs to deal with large amounts of data, she can buy more resources (more disk, more memory, more CPU). This way, the user can buy as few or as many resources as needed by the data and does not have to deal with maintaining its own computer resources, upgrading them, etc. That is, the user can concentrate on gathering and analyzing the data.

The most famous examples of cloud computing are AWS (Amazon Web Services), Microsoft Azure, and Google Cloud services. All of them offer multiple services, including relational databases. For instance, AWS offers MySQL and Postgres, as well as Amazon Aurora (a database that is MySQL and Postgres compatible), Amazon RDS, and Amazon Redshift (a data warehouse). Microsoft Azure offers Azure SQL database, Azure database for MySQL, Azure database for PostgreSQL, and SQL Server. Google Cloud offers Cloud SQL for Postgres, Cloud SQL for MySQL and Cloud Spanner. As it can easily be seen, all the main providers support MySQL and Postgres, besides their own offerings. All that has been learned in this book can be ported to this new environment.

As a simple example, assume we have created an account with AWS to manage a MySQL database. The only extra step needed is to create an AWS *instance*. This is simply a specification of what services we want to use and how many resources we need. AWS Cloud resources are housed in several data centers, each one in certain areas of the world (called a *region*); within each region, there are different locations (called *Availability Zones (AZ)*).[7] Because of this, the user has to start by choosing a region and an AZ for the database.[8] Once this is done, the user can create a database. A choice of databases is available; the user in this example would pick MySQL. Then the choice needs to be configured by telling AWS the level of resources needed and a few more details. Next, the user picks a database name, a username, and a password and provides AWS with some more information, this time concerning the database, not the instance (for instance, one can choose whether to encrypt communication with the database for security; whether the database should be periodically backed up; whether database access should be monitored, etc.). Once this is done, the particular database (called a *database instance*) is created. From this point on, the user can communicate with this database instance using a GUI (for instance, MySQL Workbench; see the previous Appendix) or the CLI (Command Line Interface), just like one would do with a regular database. For instance, when connecting to the database the user would enter as hostname the value provided by the AWS Console (it is called an *endpoint* there), and as username and password the values entered when creating the database instance. Now the user can do everything that she would do with a local database. To load large amounts of data, the easiest way is to use AWS DataSync, which is a service for transferring data between a user's computer (or any other data repository) and AWS computers.[9] For smaller amounts of data, one can use `INSERT INTO` statements. Data manipulation and analysis happens through SQL, and everything we have seen in this book applies.

---

[7] For instance, at the time of writing this, the USA has an East Region (with AZ in Virginia and Ohio) and a West Region (with AZ in California and Oregon). There is also a region for Asian Pacific (with multiple AZ in Japan, one in India, one in Singapore, and one in Australia), one for Europe, one for Canada, and one for South America.

[8] All this process is traditionally done through a Wed, form-based interface called the *AWS Console*, so it is a matter of clicking away (although one can also use the CLI).

[9] See https://docs.aws.amazon.com/datasync/latest/userguide/what-is-datasync.html for documentation.

References

1. Ricardo Baeza-Yates and Berthier Ribeiro-Neto. *Modern Information Retrieval.* ACM Press, 2nd edition, 2011.
2. Carlo Batini and Monica Scannapieca. *Data Quality: Concepts, methodologies and techniques.* Springer, 2006.
3. Richard Belew. *Finding Out About.* Cambridge University Press, 2008.
4. Michael Berthold, Christian Borgelt, Frank Höppner, and Frank Klawonn. *Guide To Intelligent Data Analysis.* Texts in Computer Science. Springer, 2010.
5. Tamraparni Dasu and Theodore Johnson. *Exploratory Data Mining and Data Cleaning.* John Wiley and Sons, 2003.
6. AnHai Doan, Alon Halevy, and Zachary Ives. *Principles of Data Integration.* Morgan Kaufmann, 2012.
7. Xin Luna Dong and Divesh Srivastava. Big data integration. *Synthesis Lectures on Data Management*, 7(1):1–198, 2015.
8. Joseph M. Hellerstein. Quantitative data cleaning for large databases. Technical Report, UC Berkeley, 2008. https://dsf.berkeley.edu/jmh/papers/cleaning-unece.pdf
9. Jeroen Janssens. *Data Science at the Command Line.* O'Reilly, 2015.
10. Daniel Lemire and Anna Maclachlan. Slope one predictors for online rating-based collaborative filtering. *CoRR*, abs/cs/0702144, 2007.
11. Gordon Linoff. *Data Analysis Using SQL and Excel.* Wiley, 2008.
12. Christoper D. Manning, P. Raghavan, and H. Schutze. *Introduction to Information Retrieval.* Cambridge University Press, 2008.
13. Mehryar Mohri, Afshin Rostamizadeh, and Ameet Talwalkar. *Foundations of Machine Learning.* The MIT Press, 2012.
14. Kevin Murphy. *Machine Learning: A Probabilistic Perspective.* MIT Press, 2012.
15. David Skillcorn. *Knowledge Discovery for Counterterrorism and Law Enforcement.* Chapman and Hall/CRC Data, 2008.
16. Pang-Ning Tan, Michael Steinbach, Anuj Karpatne, and Vipin Kumar. *Introduction to Data Mining.* Pearson, 2nd edition, 2019.
17. Robert Trueblood and John Lovett. *Data Mining and Statistical Analysis Using SQL.* Apress, 2001.
18. Hadley Wickham. The split-apply-combine strategy for data analysis. *Journal of Statistical Software*, 40(1), 2011.
19. Hadley Wickham. Tidy data. *Journal of Statistical Software*, 51(10), 2014.
20. Hadley Wickham and Garret Grolemund. *R for Data Science.* O'Reilly, 2017.

A. Badia, *SQL for Data Science*, Data-Centric Systems and Applications,
https://doi.org/10.1007/978-3-030-57592-2

Index

A. Badia, *SQL for Data Science*, Data-Centric Systems and Applications,
https://doi.org/10.1007/978-3-030-57592-2

Printed in the United States
By Bookmasters

Printed in the United States
By Bookmasters